Sheet Metal Fabrication *Techniques and Tips for Beginners and Pros*

Sheet Metal Fabrication *Techniques and Tips for Beginners and Pros*

Eddie Paul

motorbooks

First published in 2008 by Motorbooks, an imprint of MBI Publishing Company, 400 First Avenue North, Suite 300, Minneapolis, MN 55401 USA

Motorbooks titles are also available at discounts in bulk quantity for industrial or sales-promotional use. For details write to Special Sales Manager at MBI Publishing Company, 400 First Avenue North, Suite 300, Minneapolis, MN 55401 USA.

To find out more about our books, join us online at www.motorbooks.com.

ISBN-13: 978-0-7603-2794-4

On the cover: Making a flared fender requires knowledge of many aspects of sheet metal fabrication.

On the back cover: A small finger brake is useful for bending flanges.

About The Author
Beginning as a self-taught welder, painter, metal fabricator, and machinist, Eddie Paul has gone from customizer to creator, designer to inventor. He has built all manner of outrageous mechanical contraptions for the entertainment industry as well as more than 50 cars for movies like *Grease, ET*, and *The Fast and the Furious*. His company, which performs an array of engineering, design, and manufacturing services, has served clients as diverse as the Department of Defense, Boeing, and Rockwell. He lives in El Segundo, California.

Editor: James Manning Michels
Designer: Christopher Fayers

Printed in China

CONTENTS

Introduction 6

CHAPTER 1: **Metal Fundamentals** 7

CHAPTER 2: **Pattern Making with Paper, Cardboard, Plastic, and Wood** 17

CHAPTER 3: **Hammer Forming** 35

CHAPTER 4: **Metalworking Tools and How To Use Them** 46

CHAPTER 5: **Big Equipment for Metalwork** 74

CHAPTER 6: **Cut, Bend, and Shape** 77

CHAPTER 7: **Metal Finishing** 89

CHAPTER 8: **Metal Treatment** 91

CHAPTER 9: **Roll Bar Fabrication** 100

CHAPTER 10: **Sheet Metal Projects** 108

CHAPTER 11: **Training Aids** 156

CHAPTER 12: **Manufacturer Source Guide** 157

Index 159

INTRODUCTION

I recently attended a tool show and went over to the body and fender section to be "blown away" by the new body and fender hammers. Who would ever have thought they could make a body hammer with a small red stripe around its wooden handle, or that the metal heads could be chrome plated? Hey! Look at the new dollies—they now come in a cardboard box and, boy, are they shiny!

I am, of course, using absurdity to make a point. The point is we have gone from horses and buggies to rockets, from vacuum tubes to microchips, and from walking around looking up at the sky to flying a rocket into outer space. However, during this technological evolution, the methods and tools used for shaping metal have changed little. To repair a dent, you still use a hammer on one side of the metal, hold a dolly on the back side, and tap until the dent is gone. Or, if you are forming a new piece of metal for a project, you still tap away until the shape is the way you want it to be. Forming metal this way takes many hours, but it is still the way everyone does it.

Metal shaping is a long and laborious task involving a lot of old technology. The skilled craftsman has learned how to swing the hammer and hold the dolly to get the results they need. This art has been handed down through many generations of metal smiths.

Take my family. My grandfather worked as a fabricator of locomotive boilers for the railroads. One of my uncles helped build giant steel submarines during the second World War, and another worked building war ships. My dad owned and operated a few companies, one of which was a sheet metal shop, another a thread rolling business, and a third that manufactured pipe. When my mom was young, she helped build aircraft. So, you might say metal is in my blood. In fact, on my mom's side of the family we have the Studebakers. Yep, those Studebakers. My mom's father was a top-notch machinist who built and used his own lathes and mills; her brother is an automotive writer. For the last 40 years, I've earned a living working with metal.

The plain and simple fact of the matter is that since even before my grandfather's time, the art of metal working has changed little.

I, and many others, have invented and built lots of tools that make metal shaping easier, but there has been no real change in the process of shaping metal. Metal is one of those things that has gone about as far up the evolutionary latter as it can, and it still involves a lot of labor. The very fact that it is not a lazy person's occupation is what attracts me to it.

Unlike things made of plastic or wood, an item made out of metal demands respect. Steel represents strength and power, and forming steel represents control over that strength and power.

The ability to make metal bend or grow or even be reduced in size or do what you want it to do, gives you a feeling of power and satisfaction like nothing else. Metal, for all intents and purposes, is eternal. Wood, on the other hand, will be eaten by termites or rotted by the sun within your lifetime. Most covered wagons are long gone, their wooden structures having rotted away in the desert; only their steel rims and other metal parts survive. Metal will not last forever, but most of it will outlast us. Anyone with a circular saw and a hammer can make a table, but to heat and form metal peices, weld them together into one, and grind it to a smooth, flat finish, now that takes skill . . . and of course a lot of really, really old tools—old, rusty tools made out of metal.

CHAPTER 1
METAL FUNDAMENTALS

The practical and creative applications for custom metal fabrication are virtually unlimited. So, too, are the techniques, skills, tools, and machines that are all part of the metalworking process.

Metalworking is all about creating new metal objects and modifying existing metal objects. There are literally hundreds of different metals and metal alloys, and an equally vast selection of shapes (such as flat plate, round and rectangular bar, or tubing) and stock (such as the alloy or type of metal—i.e., steel or aluminum). The types of metals we'll be working with throughout this book are aluminum and aluminum alloys, mild, cold-rolled, hot-rolled, sheet, bar stock, tube, and angled steel. For automotive fabricators, customizers, and body repairmen, learning how to work these materials will allow you to handle just about any job that comes your way.

Successful metal fabrication requires that you thoroughly understand the basic properties of the metal you'll be working with and that you have a practical working knowledge of the tools and equipment you'll be using. We'll get into metalworking tools and machines a little later. First we'll look at metals and their properties.

Did you know that there are more than two dozen different types of aluminum alloys alone? Each and every one possesses unique properties that makes it perfect for some applications and unusable for others. Some are different enough that they cannot be welded together; others may break when subjected to a particular force. For these reasons, any good text on customizing and fabricating must start with a lesson on the science of metal, otherwise known as *metallurgy*.

When it comes to cars, trucks, and motorcycles, mild steel is most commonly used for modification and fabrication, since it can be fused directly onto a steel body or frame through welding. All other materials, including fiberglass, plastics, and aluminum, are more often relegated to bolt-on status, although they can be attached and molded to steel with some degree of longevity through chemical bonding. You may notice the occasional professional fabricator who elects to fabricate with an aluminum alloy or even with stainless steel. There are certain advantages to working with these metals in terms of quality and finish; however, I would highly recommend sticking with plain mild steel during your basic learning process.

Stretching or shortening a frame is easy if you have some basic tools and some basic knowledge.

Working in metal requires a few easy-to-learn skills including welding and cutting. In the past few years, cutting has evolved into little more than setting a dial and following a straight line with your plasma gun.

As I said, metallurgy is simply the science of metal and metal alloy. Everything from selecting the correct type, composition, gauge (thickness), and size of metal to welding, treatments, and metalworking techniques will be better understood once you have a fundamental knowledge of metallurgy. An experienced metalworker can quickly determine the type of metal with which to fabricate a roll bar, or what alloy to use for the aluminum panels in an engine compartment.

In this chapter, we'll cover the fundamentals of metal. The basics, rather than the heavy science of metallurgy, are all you need to know in order to become proficient at working and fabricating with metal.

Once you become familiar with the different alloys, their recommended applications, and the best ways in which to work with them, you will find that you can easily determine the specific types of metal and fabricating techniques that will help your project along. You'll also know exactly what to stock in your shop's supply of metals so you will already have what you need when it comes time to fabricate that special part.

The two types of metal that you will be working with most frequently on automobiles and motorcycles are steel (mild and, to a lesser degree, stainless) and aluminum. The differences between the various types of steel and aluminum come from their extra ingredients, or alloys. These are, essentially, just other types of metal blended together with the base metal to enhance one or more aspects, such as strength, corrosion resistance, ductility, or malleability. The weldability of metal can also be changed just by adding a percentage of one or more metals.

The first thing to remember is that a metal's alloy content is an important factor to consider during the working stages and, of course, for the structural integrity of its application in the first place. Here in my customizing shop, Customs by Eddie Paul, we often use 3003-H14 aluminum alloy (refer to the alloy definitions below) for much of the fabrication that we do. The 3003-H14 has superior strength characteristics over pure aluminum, is easily welded with either tungsten-inert-gas (TIG) or oxygen-acetylene gas welders, yet remains malleable for shaping and bending. By comparison, a 6061-T6 aluminum alloy would yield even more strength than the 3003-H14, but the 6061-T6 is also more brittle and, if welded, may develop stress cracks at the weld.

SHEET METAL QUALITIES

Before starting to shape your part, you must decide what the most important qualities are for that part. Should it be strong and lightweight, or is appearance more important? There are several varieties of aluminum and steel sheet metal available, and this section will address the differences.

Stress is the load in pounds for every square inch of cross sectional area. This is expressed as pounds per square inch, or PSI. For example: if a one-inch-square bar is pulled by a 1000-pound load, the bar is stressed to 1000 pounds divided by one square inch, equaling 1000 pounds stress. Halving the load cuts the stress in half. But halving the area doubles the stress.

Yield point is the point after which metal will not return to its original shape when the load is removed.

Tensile strength is the maximum stress a metal can handle before it fails.

TYPES OF ALUMINUM ALLOYS

The following is a list of aluminum alloys defined by four-digit numeric codes that identify the alloy content. Specific commercial alloys have a four-digit designation according to the international specifications for wrought alloys, or the ISO alpha-numeric system. The first digit represents the main element of the alloy. The alphanumeric code that follows the four digits (i.e., "H14" or "T6") is the hardness and temper specification of an alloy. For example, a letter "F" in the temper code refers to *fabricated*, which is an aluminum that has not been treated for hardness. A letter "O" indicates *annealed*, or softened by a process of heating and cooling. A letter "H" indicates a *strain-hardened* alloy (hardened by cold-working), and a letter "T" means *heat-treated*. Generally speaking, the higher the number in the temper code, the harder and stronger the alloy.

- 1XXX (1000-series) is the designation for mostly unalloyed (99% pure) aluminum. Alloys in the 1000-series offer high corrosion resistance, excellent workability, and easy weldability; however, their low tensile strength precludes use in certain applications. For instance, if used in sheet form, as in the body metal for a car or motorcycle, 1000-series is great for forming using a sandbag and hammer and will become "work hardened" during the hammering. In fact, this is the alloy of choice for most metalworkers if aluminum is the metal they are using for a project. This is a common alloy for use in automotive fabrication where strength is not an issue. Not heat treatable.
- 2XXX (2000-series) is an aluminum containing copper as its main alloy. 2000-series aluminum alloys provide a better strength-to-weight ratio than 1000-series alloys, and they are also easy to work with. The trade-off, though, is that these alloys are not as ductile, meaning that bend radii must be fairly large and gradual, and joining pieces of 2000-series alloy must be accomplished by riveting or chemical bonding rather than welding due to some of the alloying elements, like copper. Welding aluminum of these types will not only hurt the mechanical properties, but will also destroy the corrosion resistance, practically destroying the properties that made the metals useful in the first place: therefore they are considered non-weldable, except by resistance welding. Heat treatable.
- 3XXX (3000-series) indicates an aluminum with a main alloy of manganese. The addition of manganese yields a 20% increase in strength over 1000-series alloys, yet it retains the working qualities of pure aluminum and can be TIG or gas welded. For these reasons, 3000-series aluminum alloys are the most popular choice among automotive fabricators. Not heat treatable.
- 4XXX (4000-series) is an aluminum alloyed with silicon. Moderate strength. Not heat treatable
- 5XXX (5000-series) is an aluminum alloyed with magnesium. Moderate-to-high strength. This alloy can also be welded using MIG, TIG, or gas. Not heat treatable
- 6XXX (6000-series), such as 6061-T4 or 6061-T6, is commonly used in production due to its relatively low cost and excellent mechanical properties. Annealed 6000-series aluminum alloy (or 6000-series with an "O" temper code) also lends itself to forming. Weldable. Heat treatable.
- 7XXX (7000-series) is an aluminum alloyed with zinc. 7000-series offers the greatest strength but is the least ductile. Not recommended for welding. Heat treatable.

Usually we repair dents and try to get a panel as straight as possible. Occasionally, however, we are asked to make a car or truck look damaged. Using a rosebud heating torch comes in handy for deforming metal and knowing which way it will warp once heated.

Although aluminum sheet metal comes in a variety of grades, there are four commonly used types:

1100-series aluminum is pure, commercial-grade aluminum. It is soft and work-hardens slower than other grades. It is also the best grade for welding.

2024-series aluminum is mixed with copper during manufacturing. It is stronger and more fatigue resistant but cannot be welded. This type is primarily used for manufacturing airplane parts.

3003-series aluminum is mixed with manganese during the manufacturing process. It is stronger than 1100-series aluminum but retains the same shaping properties. This grade is most popular for bodywork on cars.

6061-series aluminum is pure aluminum with small amounts of manganese or magnesium/silicon added during manufacturing. This is the least expensive and most versatile of the aluminums previously mentioned.

TYPES OF STEEL

Fortunately, the selection of steel that can be used for custom metalwork and fabrication is more abbreviated than that of aluminum, and therefore less confusing.

Steel is an iron alloy. There are two types: *carbon steel* and *alloy steel*. While some advanced metalworkers and high-end customizers make liberal use of *stainless* steel, the level of technique required to work with it is likewise at the higher end of the scale. Stainless steel is a corrosion-resistant steel commonly alloyed with a high percentage of chromium and nickel. By the way, not all stainless steel is "stainless." Like aluminum, there are several stainless

Above and below: On the Cobra movie car, patch panels and replacement panels were required. When parts for older cars are just not available, the skill to make what you need is a basic requirement.

alloys with varying degrees of corrosion resistance, strength, etc.

For a detailed description of metals and their characteristics, I highly recommend that you obtain a catalog from a good metal supply company. Such catalogs often have a sizeable reference section that provides useful information on all metals and alloy content. There are many appealing structural and cosmetic qualities associated with the use of stainless steel, so you may want to consider advancing your skills once you have mastered the basics presented in this book.

For example, occasionally there are fabrication jobs that require the strength and weight of steel along with the corrosion resistance of aluminum. For example, the 14-foot mechanical great white sharks that I built for the Cousteau's Discovery Channel specials are framed entirely out of stainless steel. With constant exposure to the harsh salt water, any part of the shark structure made of carbon steel would have corroded and failed within a few short days.

For general automotive work, my use of stainless steel is usually limited to hardware items like fasteners (bolts, nuts, washers, etc.). Occasionally, a job comes up where we need to use stainless to fabricate portions of a frame or some brackets. Stainless can be very easily machine polished to a high chrome-like luster. But the cost factor for both material and labor usually keeps us working with carbon steel.

Carbon steel, a combination of iron and carbon, is the type of steel that will be used in most of the techniques in this book. But to avoid any confusion down the line, there are a few other terms that I may use in reference to steel.

One is *mild* steel. Mild steel is simply carbon steel that contains a maximum of 0.20 percent carbon. Mild steel cannot be hardened or tempered, but it can be casehardened, and it is easily weldable using MIG, TIG, or gas welding.

Hot-rolled steel is carbon steel that is brought up to a white heat during its manufacture and then passed through a series of rollers to reduce the cross section (make it thinner), thereby increasing its length. It is then cooled, cut to length, or coiled. It is also very easy to weld.

Cold-rolled steel is carbon steel that is manufactured by a process technically referred to as *cold reduction*. The cold-reduction process reduces, as its name implies, the thickness of steel by rolling or drawing the material without preheating it. This cold method adds strength as well as producing stock that is smoother and more consistent. It can be welded by any method.

The process of hot rolling produces a surface slag that, when compared side-by-side with cold-rolled steel, is quite obvious. The benefit to using hot-rolled steel is lower cost. The more-expensive cold-rolled steel is commonly used in precision sheet metal applications, since it provides an excellent surface, better material consistency, and a more accurate thickness.

The same basic code system that defines aluminum alloys similarly defines steel. But before we get into coding, let me say that I seldom order my steel by code, as I do with aluminum. The main reason is that I've developed a rapport with my regular metal supplier, and I simply refer to my carbon steel orders as either *hot-rolled* or *cold-rolled.*

When it comes time for you to locate a metal supplier and place an order, remember that a good supplier will have a product catalog that also contains a wealth of useful information pertaining to sizes, gauges, and alloys. A knowledgeable salesman will also take the time to help you with your order based on your specific requirements. Still, it's always good to know what you're ordering. This is not a complete list of codes; I've narrowed it down to the basics to keep things simple for now. All can be welded.

- 1XXX (1000-series) Basic open-hearth and acid Bessemer carbon steel that is non-sulfurized. 1020-series cold-rolled steel sheet metal is a common material for automotive fabrication.
- 2XXX (2000-series) Steel alloyed with nickel.
- 3XXX (3000-series) Steel alloyed with nickel and about 1.25 to 3.50 percent chromium.
- 4XXX (4000-series) Steel alloyed with molybdenum or nickel-chromium-molybdenum. You've probably heard the term "4130 chrome-moly" a few times. 4130 is steel alloyed with chromium and molybdenum. Stress-relieved 4130 chrome-moly is used where structure strength is most critical. Annealed chrome-moly is used when fabrication requires forming and bending.

The code series for steel continues up to 9XXX (9000-series), with different alloys and percentages of additional metals added to enhance different features and characteristics of the base carbon steel. As you get more involved in metalworking, you'll start to recognize specific types of steel whose properties lend themselves to your particular forming process.

My steel preference for general all-around metal fabrication are two alloys referred to as AK and SK steel sheet ("A" indicates the addition of aluminum, "S" indicates the addition of silicon; both are added during the killing process, indicated by the "K"). My regular metal supplier, M&K Metal in Gardena, California, has both AK and SK steel sheet stock and will sell single and even partial sheets, whereas some metal suppliers will only sell these types of metals in mega-pound quantities.

You will find these metals to be the best all-around alloy for sheet metal fabrication of parts and panels. You

The ability to weld all types of metal is a basic skill the metal fabricator needs, and a welding machine is his basic tool.

will notice that when you work AK or SK steel it will not work-harden as quickly as regular cold-rolled steel does. This is a very big advantage when fabricating deeply contoured panels. If you cannot find AK or SK at your metal supply company, try calling a local customizer or stamping company—they might be willing to sell a few sheets to a fellow fabricator. I have, on many occasions, gotten together with someone else and placed a combined order; this can even lead to a quantity discount.

Most of the AK sheet metal that I use for the metalworking demonstrations in this book is 18-gauge. When it comes to working with car bodies, I try to match the gauge of the panels already on the car. For stand-alone projects, 18-gauge is a little heavier than necessary, but it does allow for deeper shapes to be formed into the metal. Although 20-gauge would be easier to cut and shape, 18-gauge sheet is perhaps the best for a beginning fabricator.

THE THICKNESS OF METAL

The gauge of sheet metal is a numeric reference that indicates its thickness. It's similar to the scale for electrical wire in that a numerically higher gauge indicates a thinner material. I suppose you can refer to this as "the inverse law of logic as it pertains to sheet metal gauge." The best way to determine the precise gauge of a sheet of metal, short of measuring with a set of calipers, is to use a handy little sheet metal gauge tool.

In a side-by-side comparison, the same gauge number of a sheet of steel (ferrous) and a sheet of aluminum (non-ferrous) is different in actual thickness; in other words,

A decent-quality tube bender and the knowledge of how to use it will allow you to make your own frames, cages, and roll-bar hoops.

the two sheet materials with equal measurements in thousandths of an inch will have different gauge numbers. For example, 20-gauge steel is 0.0359" thick while 20-gauge aluminum is 0.0320" thick, not a big difference, but enough to be confusing to some of the engineers out there. So 20-gauge aluminum is closer to 21-gauge steel, which is 0.0329" thick. For those of you who will work primarily with sheet steel, this is something to keep in mind when placing the odd order for aluminum sheet.

METAL SHRINKING: HOW TO SHRINK METAL AND WHY

So what exactly is metal *shrinking*? Well, to a fabricator, it's when you literally pull or press a section of metal together into itself. Doing this doesn't actually make any metal go away, but it reshapes it and compresses a particular section of metal. This is one way to form curves in an otherwise flat or straight piece of metal.

Now you're probably wondering *how* do you shrink metal? There are a number of ways to accomplish this. One of the basic methods used by autobody repairmen is to use a pick hammer with a padded dolly and apply small indentations into a section of metal. These small dents will have the effect of shrinking a section of metal by stretching it into a smaller area!

Confused? Well, imagine a rock that hits the center of a car door and puts a hemispherical dent in the metal. That rock has just stretched the door metal at the area of impact, but the surrounding area has been pulled toward

the impact, therefore shrinking the door skin overall.

On the other hand, if you use a pick hammer (which is a small-bodied hammer with a pointed end) and back the panel with a padded or rubber-coated dolly (or a block of wood), then, as you pick the panel, you are basically doing the same thing to your metal sheet as the rock did to the car door, only with a controlled technique applied to a specific area. The reason for using a padded dolly or block of wood instead of a steel dolly is that, if you used metal for a backing, you would actually be stretching the metal, not shrinking it. Pounding metal between a hammer and a steel dolly tends to thin the metal and, since metal has to go somewhere, it expands outward into the surrounding metal. By using a soft block and a pick hammer, you allow the metal to form small peaks and thereby pull the outer metal toward the small peaks.

I know I'm oversimplifying the process, but doing so makes it easier to understand the process of shrinking, and the more you understand the better you can work the metal. Just remember that for every action there is an equal and opposite reaction, so moving metal from one spot will cause it to go somewhere else. The trick is to know where the reaction will occur and in which direction the metal will react. Then and only then will you become master of the metal.

The process of shrinking metal with the use of a torch is well known and pretty standard. The accepted method is to pinpoint the area that requires shrinking, which would be a high spot in an otherwise smooth panel, then heat a small spot about the size of a silver dollar. Then as it turns cherry red and raises to form a small bump, simply tap the bump down slightly until the spot is level with the surrounding metal.

Before the metal has a chance to cool, you can quench the spot with water to accelerate the process; however, this tends to harden the metal and make it more difficult to work with later. If you simply allow the metal to cool off naturally, the metal can be worked without having to anneal it again. Hot-shrinking techniques may vary from one fabricator to the next, but the best way to know how this process works is to experiment with a scrap sheet of metal.

STRETCHING METAL

Stretching metal is the opposite of shrinking and is the most common mistake that novice metalworkers make when working, or, more accurately, overworking, a panel. It is the hammer-on-dolly work that thins and stretches the sheet of steel that requires shrinking it into its proper shape. Remember, you can shrink metal as much as you

The bottom hinge area on the right door frame was missing and had to be refabricated from scratch. As you can see, understanding metal is a very basic requirement.

want (to a point), and it will only get thicker; but if you stretch it too much, it will eventually tear when it becomes too thin to have any tensile strength.

Whether shrinking or stretching a sheet of metal, you'll notice (if you're observant) that the worked area actually hardens. Why does the metal get hard after shrinking or stretching? We call this "work hardening," and it's the direct result of squeezing the metal's molecules so close together that the metal gets tougher and harder to work. If you run into this while working with metal, and I'm sure you will, you can simply anneal the metal to soften it again.

The thing to remember when you first start working with metal is that shrinking will thicken the metal, which can be stretched later, but stretching the metal will decrease the thickness and make it harder to work with if you need to shrink it later. So if in doubt, shrink it first. Or as I like to say, err on the side of thicker.

HOW TO HARDEN AND SOFTEN METAL

The simple process of hammering metal compresses the molecules and hardens the metal. Once this happens, the equally simple process of heating metal with a torch and letting it cool slowly (called *annealing)* lets the molecules separate back into their own original space, softening the metal and making it workable once again. This is an over-simplification, but it is the easiest way to describe a fairly complex process.

There is another way to harden metal, and that is to heat it up and then bring the temperature back down very

quickly by dunking it in oil or water. So, simply by differentiating the metal's temperature over a frame of time, you can control the overall hardness of a sheet of metal, making it workable or not.

For example, annealing is easily performed with an oxygen-acetylene gas torch by heating the metal to its critical temperature (approximately 1200 degrees Fahrenheit) and then allowing it to slowly cool. The process of *tempering* (to make it harder and tougher) requires heating the metal to its critical point and then quenching it in water or oil.

ANNEALING SHEET STEEL

When working with steel, you will find that most types are easily worked, especially if you are using AK steel (available at most large metal yards). But in certain instances you will find that you have overworked an area with too much hammer and dolly work, making it "work hardened" to the point where it no longer responds to light taps of the hammer. It is then that you will have to resort to re-softening (annealing) the area you are working.

To anneal sheet steel, you will have to heat it until it's cherry red and then allow it to cool slowly. This will take a bit of time, and you should experiment on some scrap sheet metal to get the hang of it. Start off by heating the area that you need to work. Keep the working area as small as possible. Work slowly and keep the torch moving around. Use a rosebud tip that gives a wide flame that covers a broad area. A standard welding tip can work if

This is a '65 Mustang I built many years ago. The rear flares were our area of expertise; we were the only company that made them out of steel, and we had a waiting list of customers for them. Each flare took one week to make.

you're careful, but it tends to concentrate the heat of the flame in too small an area for good results.

Begin by bringing the temperature of the metal to slightly above its critical point, which varies for different types of steel. Low-carbon steel should be annealed at about 1650 degrees F and high-carbon steel at between 1400 and 1500 degrees F. Constantly move the torch, maintaining the critical point temperature just long enough to heat the entire piece evenly throughout. This will be indicated by the color of the metal, which should be a uniform bright red. If you go above the critical point temperature, the metal in that area will melt.

Also, if steel is heated too much above the critical point (but doesn't melt) then, following proper steel annealing procedure, is cooled slowly, the coarseness of the grain will fix itself at the coarseness it held at the maximum temperature. In other words, the grain of improperly annealed steel is coarser. You will want to avoid this, since it will change the working attributes of the metal in that area, possibly causing cracks.

ANNEALING SHEET ALUMINUM

Aluminum is annealed just the opposite of steel–whereas steel is heated then cooled slowly, aluminum is heated and cooled quickly, either by dipping it in water or by blowing it with a blast of air.

It is a good idea to anneal aluminum prior to working it, because this makes it far easier to form the metal. As you form the metal, you will find it getting progressively harder and more difficult to shape. This is referred to as "work hardening," and is a result of the molecules in the metal being compressed together by repeated hammering or rolling in the English wheel. In either process, as the metal molecules are compressed together, the metal actually becomes denser. Annealing the metal will allow the molecules to decompress, thereby softening the metal.

You will need to acquire some additional tools, such as a torch holder that can hold the oxy-acetylene torch head while you move the pieces of metal around. This is not an absolute necessity, since you can simply turn the torch off and set it aside while you work the metal. Another tool is an annealing plate. This can be as simple as placing the metal on a homemade grate over an empty metal trashcan.

The first step of the annealing process is to place the metal that needs to be annealed on the grate. Then, set the torch to an "acetylene-rich" setting by turning down the oxygen and lightly brush over the aluminum with the flame. This will deposit black soot over the aluminum. The coating of soot will serve as a temperature gauge when you start annealing the aluminum because it will take about 700 degrees from a neutral flame to burn off the soot. This is the ideal temperature required to anneal the aluminum. If heat in excess of 700 degrees is applied during the annealing process, you will melt the aluminum.

Now adjust the torch to a neutral flame and burn off the soot, moving the torch so you do not apply too much heat in one spot and melt the metal. Slow circular paths will control the heat and burn off the soot.

If 2024 aluminum is used, it has to be annealed in a very controlled process, all but impossible for the small builder to perform. Better choices would be aluminum alloys like 1100 and 3003, which can easily be annealed by heating with a torch and cooling with water or blown air. However, since you need to heat these alloys to near their melting temperature, around 1200 degrees Fahrenheit, it is essential to monitor the temperature.

There are a few ways to do this, some good and some not so good. The simplest and most inexpensive is to take your oxy-acetylene torch and set the flame to pure acetylene. Now dust the metal with the flame, leaving the surface covered in the soot that the acetylene-rich flame produces. Painting of the surface this way allows you to both determine the area you are heating and the temperature to which you are heating it when you move on to the annealing process, since the soot will burn off at just the right temperature for annealing. But before you ruin a new sheet of aluminum that was just cut to size, practice on some scrap to get the feel for this method of annealing.

I said there were other ways to monitor the temperature, one of which is to buy a set of temperature crayons. Many companies sell these on the web, or you can get them at your local welding shop. Now, how not to do it. Even though an infrared thermometer may look really cool, the reflection of the metal will throw the sensor off so the reading will not be accurate and you will wind up over- or under-heating the area.

Aluminum is hardened in a totally different way from steel and is annealed differently, too. To harden steel, you heat it and quench it quickly. But that is how you *anneal* aluminum! Tempered aluminum's spring comes mostly from its alloying elements. After annealing the aluminum, the hardness will gradually return of its own accord in time.

Hardened aluminum can be temporarily annealed by heating it to about 500 degrees F and quickly quenching it with cold water. It will then remain soft for a few hours or a couple of days (depending on how hot the weather is); and it will gradually regain its original hardness. (To speed up the re-hardening process, you can re-heat the aluminum in boiling water or a 250-degree oven for a half hour or so.)

When you are done with the heating, the material can be quenched in water, and this will be determined by the alloy. (Check with the metal supply company on the particular metal you are using to find the proper method of annealing. Some aluminum alloys do not even heat-treat.) The area should now be annealed, and simply bending the edge of the sheet will prove this. From this point on, just about every metalworking process we discuss will involve the use of tools, the focus of the next chapter.

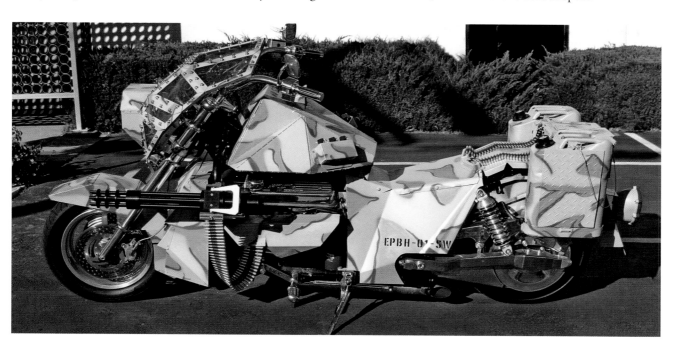

This is a Boss Hoss that I customized for a car show in just four days. It was a simple job of making the parts out of cardboard and transferring the patterns onto metal.

1 For annealing aluminum, you will need to set the torch to an acetylene-rich setting so it will produce the black soot required to "paint" the aluminum. **2** Then you simply "fan" the aluminum sheet's surface with the torch, covering it with black soot.

3 This black soot will be your temperature indicator for the annealing process. **4** Once the aluminum is covered in soot, reset the torch to a neutral flame by adding oxygen to the mixture. The flame can then be used to burn off the soot. Be careful not to hold the flame in one spot too long, or you can easily burn through the metal.

5 When all the soot is burned off the surface, the metal is at the proper temperature and can be sprayed with water or dunked in a water tank to complete the annealing process. The annealing process can be repeated as many times as needed during the metal shaping process.

CHAPTER 2
PATTERN MAKING WITH PAPER, CARDBOARD, PLASTIC, AND WOOD

Since it is much easier to cut paper, cardboard, or even wood than it is to cut metal, and since the cost of these materials is a fraction of what metal costs, it only makes sense that we make flat patterns out of paper or another material that's softer and cheaper than metal. Once the pattern is shaped and trimmed to the approximate shape of the piece to be fabricated, that design can then be traced or otherwise copied to a flat piece of metal.

Most people are in such a hurry to start the job of making a metal panel that they forgo the pattern-making process and wind up scrapping an otherwise good piece of metal or two before falling back to the pattern-making process. Others think that it may not look professional if they start with a piece of paper to institute the process of laying out a metal part. But it is, in fact, what the professionals do.

Some of these patterns are saved for future jobs. My dad had patterns he made from light-gauge sheet metal. I always thought it odd that he made a paper pattern to make a part and then, if he liked the part, he made a light-gauge metal pattern from the paper pattern just to save for future use, when he had the paper pattern already.

One day I found out why he took that extra step. I wanted to use an old paper pattern I saved from a job that I had done a few years before, just to find that the paper had gotten wet and was now worthless. It was a time-consuming lesson, so now I make metal patterns as well.

ACCUMULATION OF ERROR

Now a word about accumulation of error in pattern making. If you make a pattern, you need to mark it as the original, then use only that pattern. The following is an example of why this is important.

Say you need fifty bars that are exactly twelve inches long and you are using a cut-off wheel, or a chop saw, or even a band saw as a cutting device. There are a number of ways of performing this task, some efficient and accurate, some neither efficient nor accurate, and I have seen all types utilized even by the best fabricators.

First the bad example and the reason it is bad. A guy takes a bar of metal and measures off one foot and marks the bar, then cuts that one-foot section off the total length of the bar. He now measures and cuts off another foot-long section. He repeats the process 49 more times. The bars that he cuts are accurate (assuming he measures consistently), but the constant measuring of each piece has taken a lot of time, and the owner of the shop has died of old age waiting for his 50 foot-long pieces.

There are a couple of faster ways. One is to use the first piece to measure the rest of the parts, saving the time of pulling out the tape to measure and mark each part. Another way is to use an adjustable stop on the saw table. Set the stop so the distance between it and the saw blade is the desired length, place the stock on the table, move it gently against the stop, clamp the piece, and cut it.

Now I will give the example of the wrong, and often observed, method of cutting multiple bars of the same length. A guy (who obviously works for another shop) takes a tape measure out and measures a one-foot section and marks it. So far so good. After he cuts the first bar, he uses that piece to measure the next piece to be cut, saving the time of taking his tape measure out again. As we just discussed, this is an acceptable method. That is, until his next action . . . which is to use the second piece he cut to measure the third piece, then use the third piece to measure the fourth piece, repeating the process until he is finished. At which point he notices that the last piece looks long, real long . . . what just happened?

It is called accumulation of error. What happened was that each piece was hot from cutting, and with heat comes an effect called linear expansion. Not a lot, maybe $\frac{1}{16}$ of an inch. The expansion, combined with a dull felt-tip marker he was using (which added another $\frac{1}{16}$ of an inch) and added about $\frac{1}{8}$ of an inch to the second part, and since the second part was used to measure the third part . . . well, you see how a small error of $\frac{1}{8}$ of an inch over fifty parts can add up to $6\frac{1}{4}$ inches to the last bar.

These are just a couple of ways error can intrude into your work. Each one of us can probably come up with ten more ways, some that make each subsequent cut shorter, some that make them longer.

The solution is to make the first bar, then use only it to measure the rest of the bars. This will allow each successive bar to be measured to a consistent standard. Of course, if you measure and/or cut the first bar incorrectly, every subsequent bar you cut will also be off, but at least they will all be the same size. So, be very careful when cutting that first bar.

All of this also relates to patterns. Always resort to the original pattern, not a third- or fourth-generation pattern that has been copied and possibly rescaled during the process.

MOCKUPS IN WOOD, WIRE, CARDBOARD, AND PAPER

Mockups are different than flat patterns. Mockups give you the feel and look of an actual part, while a pattern is the mockup laid out flat. This is the difference between shape and form—shape is flat and form is three dimensional. Form is an apple, while shape is a flat cutout in the silhouette of an apple. If you are going to make a custom gas tank for your chopper, you may want to mock it up first by cutting the shape out of paper and taping it all together in the form of a gas tank, then placing it on your bike so it can be viewed from all sides. In many cases what looks good from one side may not look good from another side, and this is what a paper form will reveal.

As an extreme example of the previous statement, let's use a silhouette of a gas tank made from a sheet of plywood. The gas tank from the sides may look perfect, but from the front or the rear it will look very thin and flat. So this is why a 3-D form will help give a finished look to a shape. I make many of my mockups out of foam core, which is a foam-laminated cardboard that is easy to cut and work with while giving the needed strength required for larger mockups, such as complete front ends for cars. If you are careful when you make the mockups, you can carefully cut them apart and use them as patterns for making the part when the final design meets with your approval.

Mockups can be made from almost anything that will hold its shape, since it is only intended to show you what a finished form will look like. However, as I said, I often make them out of something that can also be used for the pattern, saving me a lot of duplication time.

MAKING A FLOORBOARD

1 It did not take long to figure out that this 1940 Ford Sedan required new floorboards. The first job was to remove what was left of the originals and grind down the frame so we could make some patterns for the new floorboards. **2** The interior was to be patterned using cardboard I had lying around. Although any size can be used, I recommend using many smaller pieces rather than one large piece, since smaller pieces are much easier to manage and can easily be taped together to make whatever size you ultimately require.

3 *The top of the frame was cleaned of the remaining original floorboard, giving us a solid flat surface on which to work.* **4** *To begin making my pattern, I lay a large piece of cardboard onto the frame and roughly mark the shape of the new floorboard.*

5 *After flipping the cardboard over, I start cutting off the edge, making the cardboard closer to the size that I need for a pattern. Keep in mind that the total perimeter need not be cut all at once. I often take one cut at a time, then refit the pattern onto the project, since I have found one cut will often affect the fit of the pattern.* **6** *In this shot, you can see that all I am concerned with are the sides. The front and rear will be dealt with as I progress with the pattern. I'll add extra pieces of cardboard as I need them.*

7

8

7 *It is a good idea to make notes on the pattern about direction and areas that will be cut or modified later.* **8** *Using a carpenter's square, I mark off the driveshaft well location that will be added later.*

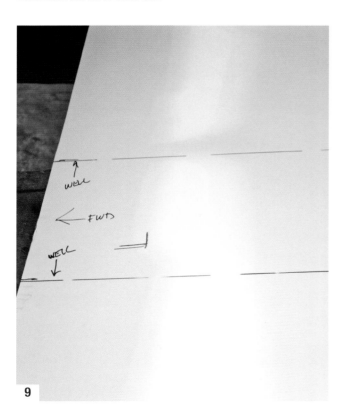

9

10

9 *As you can see, I've made notes to help guide me. For example, the big X tells me that this area will be removed later.* **10** *As I said, I will be adding to the pattern. To do this, I often just lay the next piece of cardboard on top of the first and mark the corners so it can be realigned later in the same position.*

FWD

11

12

11 *If the parts are to be realigned later, it's also a good idea to make them so that you know which part aligns with which part.* **12** *I sometimes use duct tape to keep things in position if I am the least bit worried about movement or if I know I won't have time to realign it later.*

13

14

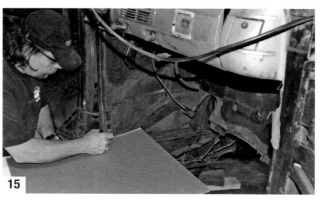

15

13 *Using the square and the frame as guides, I mark the line where the bell housing will be added later.* **14 & 15** *The opposite side's pattern is added and marked appropriately so it, too, can be cut out for the soon-to-be-built bell housing cover.*

16

17

16 *Now a large sheet of metal is laid out on a workbench along with a few notes made for the pattern that will be cut from it.* **17** *The driveshaft well is now measured and laid out on the metal in the 4-foot direction (the sheets come in 4x8- or 4x10-foot sheets). Since I did not have an 8-foot roll or brake, it made sense to make the well in two sections and weld them together later into a one-piece well.*

18

19

18 *Using handheld electric metal shears, the metal is easily cut. My dad used to use handpowered tin snips for all the cutting he did, even though electric snips were available. The word that comes to mind is* stubborn. **19** *The two sheets are cut and the edge flange marked for later bending.*

20

21

20 *I next put the pieces of metal in my slip roll and rolled them a few times to get the exact curve I wanted. I will often roll the metal a few times to get the radius instead of just rolling it once, since I do not want to overroll the metal. I found it is much easier to roll to the right curve slowly rather than try to guess the correct radius the first time.* **21** *Another trick when using a roll is to tap the starting and ending edges along their lengths with either a rubber hammer or a dead-blow hammer to get the edge to curve, since the starting and ending edges normally do not get as radiused as the center of the sheet does.* **22 & 23** *The two parts of the driveshaft well are rolled and then flanged in a brake and laid out on the pattern for a final fit.*

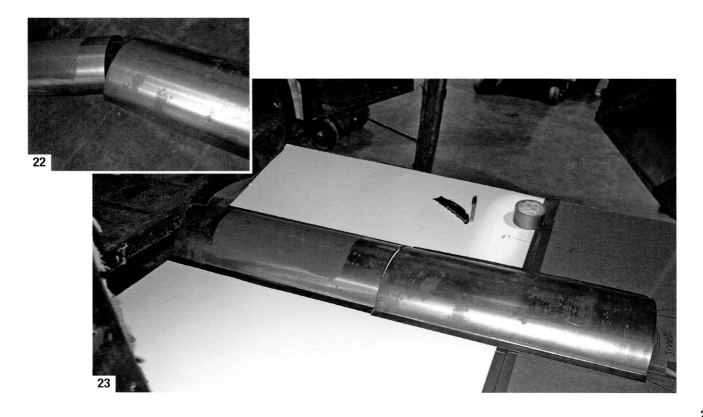

22

23

FWD ←

PATTERN MAKING WITH PAPER, CARDBOARD, PLASTIC, AND WOOD

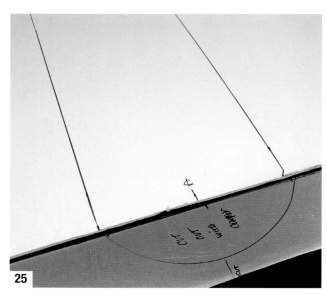

24 *The floor pan pattern is then laid on top of the metal, and the metal is marked for cutting.* **25** *The rear of the floor pans are marked for the driveshaft well cutout (the half-moon shape drawn on the metal). Also, after the cutout is made, the rest of the metal sticking out beyond the pattern will be bent upward at 90 degrees.*

26 *This photo shows one side of the floor pan bent into position and the other side before the bend is made. The driveshaft well will hold the two pieces together, as well as add a lot of strength to the whole sheet.* **27 & 28** *The driveshaft well is lined up with the two floorboards and tack welded into place.*

24

29 As the parts are welded together, the entire unit becomes stronger as a whole. The unit can now be flipped for welding the bottom with relative ease.

30 & 31 I am constantly testing the fit of the part onto the car. Every piece that is added or changed winds up changing something else, and continually fitting the part catches the warpage before it gets too far along and I have to scrap the part.

32 During one of the test fits, the front sides were a bit tight after the driveshaft well was added. It was a good time to mark the area and trim it down a bit for a perfect fit.

33

34

33, 34, 35 & 36 *After the main floor was in place, it provided enough support to use as the main pattern. The next step is to start adding to it with cardboard patterns.*

35

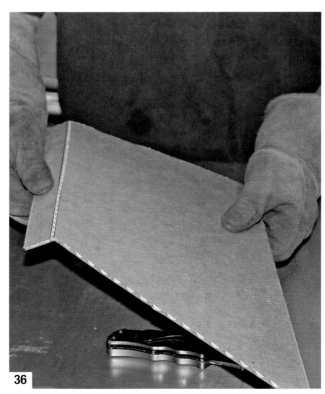

36

37 This cardboard pattern was immediately used to cut the metal piece, which was then trimmed to fit the steel floorboard.

38 Overbending the flange can be repaired by flipping the part over and tapping the back of the bend. **39** Placing the two pieces in their proper locations show the floor coming together quite quickly. What at first seemed like an overwhelming task turned out to be very simple when broken down into small, manageable jobs that concentrate on one piece at a time.

PATTERN MAKING WITH PAPER, CARDBOARD, PLASTIC, AND WOOD

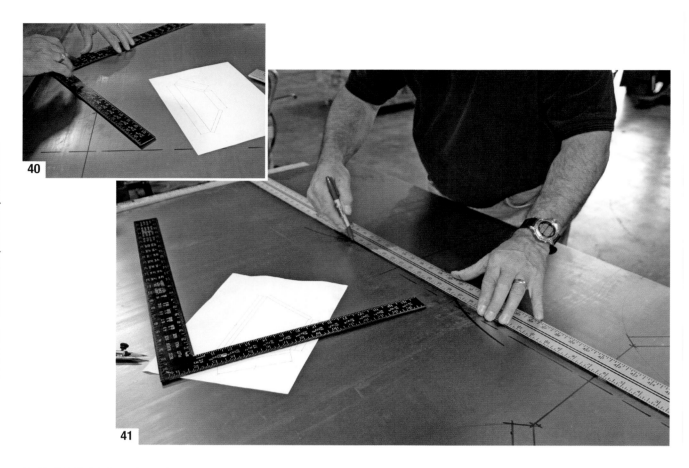

40, 41, 42 & 43. *It is time to lay out the transmission bellhousing cover, so a sketch is made and transferred to the metal, which is then cut to shape.*

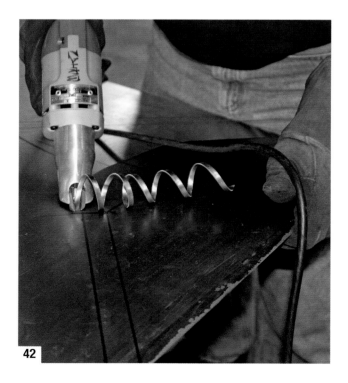

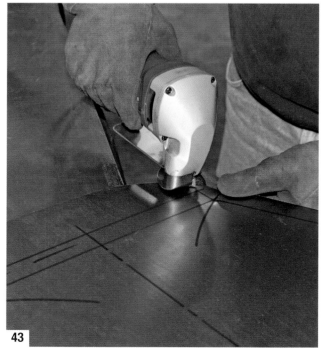

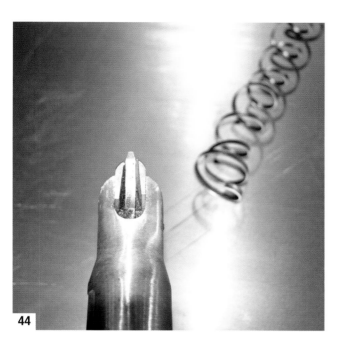

44

45

44 & 45 *Tool-talk time. The tool shown in the photo on the left is a three-blade cutter. It removes a section about ⅛ inch wide as it cuts, but it does not warp the metal as it cuts. The cutter (shear) shown on the right cuts without removing any metal, but the cut edges are not always flat—they tend to curl a bit and need to be repaired.*

46

47

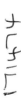

46 *The trans cover is then placed in the slip roll, and a curve is added by rolling it from the center outward in both directions to get an even curve without affecting the outer edges where the flanges will be bent up for welding.* **47** *For bending the flanges up, I use a small finger brake so I can get into the area without affecting the top and bottom flanges.*

PATTERN MAKING WITH PAPER, CARDBOARD, PLASTIC, AND WOOD

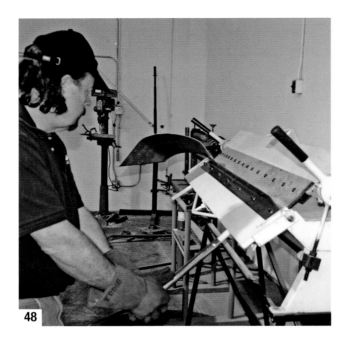

48

49

48 *I then move up to a slightly larger finger brake to get better leverage as I make sharper bends in the flanges.* **49** *The newly made part is then placed in the car and tapped into its final shape with a soft rubber hammer.*

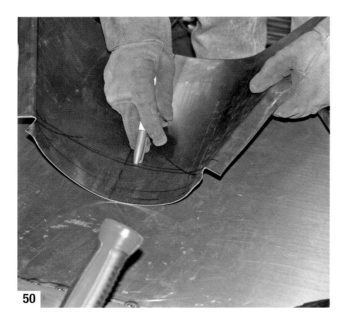

50

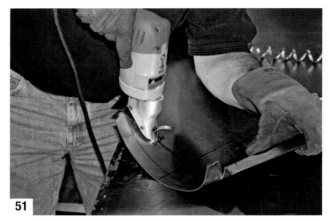

51

52

50 *Once in place, it is marked for end trimming, then removed and remarked for more accurate trimming.* **51 & 52** *The overlapping section at the rear is removed and cut to the line drawn earlier. Then it is cleaned up for refitting.*

FWD

53

54

53, 54, 55 & 56 *Removing and trimming parts and then refitting them is something you need to get used to, since you will rarely get them right the first time.*

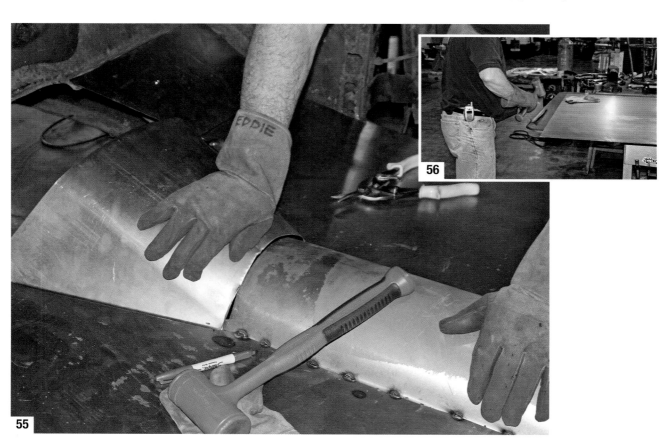

55

56

57 *Having gotten the main piece of the well close, tack weld it into place and make a filler piece.* **58** *To make the filler piece, I use a clear, thin sheet of Vivex to make a pattern. This allows me to see through it and mark my lines.*

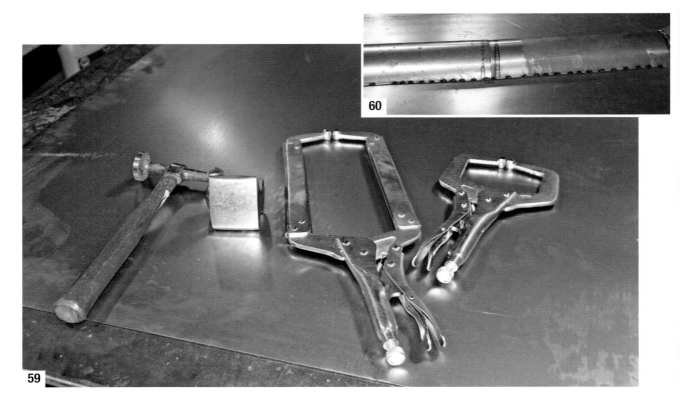

59 *Once the panel parts are tack welded into place, you will need a few clamps to hold the metal in place and a hammer and dolly to tap it flat as you weld.* **60** *That the driveshaft well is made out of two pieces poses no structural problems. It's much easier to construct with shorter pieces.*

32

FWD

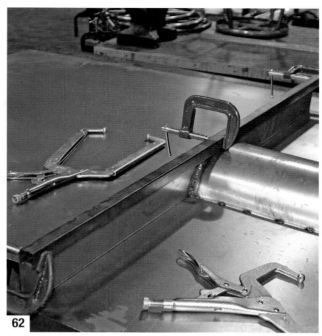

61 *As you can see, the bell housing looks good and will easily replace the original floorboard once all the welding is complete.* **62** *At the rear of the floorboard, a section of 1-inch-square tubing will add strength to the floor. It is simply clamped on and welded in place.*

63 & 64 *The entire floor can be welded together out of the car, then welded into the car at a later date when you no longer need access to the top and bottom of the floor and frame.*

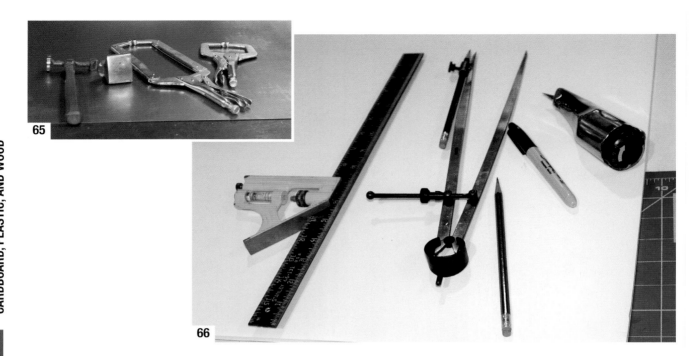

65 Tools are few for a job like this: A few hammers, a few sheet metal cutters, and a way to bend and roll the metal is all you need. **66** Most layout tools are cheap, and you will develop your own favorites. This is a small assortment of mine.

67 For measuring large areas, I made a large set of dividers out of plywood. They work better than they look. **68** I make patterns each and every day. It's a vital part of the job. Small or large, almost everything requires some sort of pattern-making skills.

CHAPTER 3
HAMMER FORMING

The art of hammer forming is a skill in and of itself. Although it may sound and look like black magic, it is in reality a very simple process if we break it down into simple steps. Like so many other skills, from afar it may look a bit overwhelming to see a finished product that has been hammer formed into a piece that looks as if it were made by a metal-stamping press at a GM plant. The concept that it was manufactured in a garage strikes the observer as impossible. But, if the process is explained well enough, the process of making a hammer-formed part becomes quite simple, as well as a handy skill to put in your arsenal of metal tricks.

To learn hammer forming, we first need to expand our vocabulary a bit. A few new words are in order; they are not large words but will be used often in this chapter and may be a bit confusing if not understood thoroughly.

A **plug** can also be called the shape, form, or 3-D pattern. But "plug" will keep the description less confusing and is the word I will be using to describe the item over which you will be shaping the metal as you hammer it into shape. The plug can be male or female; you will need to decide which type of plug you need depending on the finished form you are going to fabricate.

The male plug will allow you to shape metal over it much more easily, but the metal that gathers up will have to be shrunk to fit over the form. If the plug is female, the metal gets stretched into the concave area. It is a bit easier to pound the metal into a void with a hammer. Depending on the design of your project, it might be easier to use a male plug, or it might be easier to use a female plug. Sometimes I use both, allowing me flexibility I may need for different parts of the same design. Remember—shrinking the metal makes it thicker and stretching it makes it thinner, so metal that has been hammered into a female form can become paper-thin and even tear if the form is too deep.

Draft is the part of the shape that allows the metal to be removed from the plug once you are finished shaping it. As an example, a male plug in the shape of a cone or pyramid has draft, while a male plug in the shape of a box does not. A sphere has draft to its midpoint, after which it has what is called negative draft. The normal amount of draft that will allow a part to be removed from its plug is about 10 to 15 degrees. If you slanted the sides of a box-shaped male plug inward toward the top by 10 degrees, you would have a box with draft from which you could remove a part.

Male refers to a part that sticks out or a plug that is convex. You place and shape the metal around the outer surface of a male plug. This shrinks the metal and makes it thicker, which can lead to hardening.

Female refers to a part that sinks in or a plug that is concave. You place and shape the metal around the inner surface of a female plug. This stretches the metal and makes it thinner, which can lead to tearing.

A **matched set of dies** are used more in casting but can be fabricated for press forming on an arbor press if the required deformation is mild enough.

Jelutong is a type of pattern-making wood that is used for making plugs of all sorts. Though classified as a hardwood, it carves much like a softwood. Jelutong is easy to shape, carve, and sand to a perfect finish. It will also take paint well if the plug is to be preserved. Interesting fact: most carousel horses are made from Jelutong.

CNC router is a Computer Numerical Controlled router that describes a 3-D drawing as a series of three-space coordinates identified by ones and zeros (the computer language of binary code) to control a few stepper or servo motors working in synchronous relation to each other to create a finished part that resembles the starting art.

A **vacuum forming machine** allows you to put a piece of thermo plastic in a frame, heat it to the point of malleability, and then pull it down over a vacuum table (filled with holes connected to a vacuum pump and tank) that is supporting the plug. Then, by turning a valve, you evacuate the air from the space between the plastic and the plug so the plastic conforms perfectly to the outside of the plug. Once the plastic has cooled, it will again become rigid, and your finished part can be removed from the plug and used as a finished part.

Part is the finished product that you have made from the plug.

Shrinking metal is the process of thickening the metal by pulling it into itself mechanically or with the use of temperature control (a torch or shrinking disc) and a hammer.

Stretching metal is the process of pulling the metal apart by mechanically pulling the metal away from its center point or by pounding on the metal with a hammer on one side while holding a dolly on the other side. This flattens the metal in the area being worked, forcing it to move outward into the surrounding metal.

Annealing metal is the process of heating the metal and then letting it cool slowly or, with some metals, dipping it in water once it gets red hot. This effect will soften the metal, making it more workable.

Work hardening the metal is what happens when the metal is hammered or overworked by stretching or pounding on it too much. In many cases, the only way to continue is to re-anneal the metal.

MAKING A PLUG

The first step in hammer forming is to make a plug. In most cases, plugs are constructed from wood. Jelutong is a good choice of material because it is simple to shape. The way you shape it is up to you, according to your budget-

ary constraints. You can simply use woodworking tools like a router or a chainsaw to rough in the shape and then finish it with a sander. This is the way my dad used to make his plugs. Or, you can spend a few bucks and get a CNC router, a router that has become very affordable in the last few years.

CNC routers can cost from a few thousand to many hundreds of thousands of dollars. Today, for a few thousand dollars, you can get a good 4-foot x 4-foot CNC router that will do whatever you need. The learning curve isn't too steep, and the return on your investment is priceless. You will have traveled into the twenty-first century with the ability to make almost anything you can dream up. My router paid for itself on the first job and now is nothing but a money machine cranking out part after part for hammer forms or vacuum plugs.

Of course, you can always make your forms or plugs without a CNC router. It just takes longer.

Above left and right: *A plug is a solid piece of wood or other material over which (convex), or into which (concave), something is formed. Shown is a stack of glued-up wood waiting to be turned into a plug. This was one of the first steps in shaping the hubcaps for NBC Universal's Scooby Doo van.*

Jay Leno brought us a foam plug his guys had built while designing an aluminum intake for his Eco-Jet. He wanted us to digitize it and make a solid wood plug. In other words, we needed to extrapolate the Cartesian coordinates that described the shape, then plug the numbers into our CNC router and carve it.

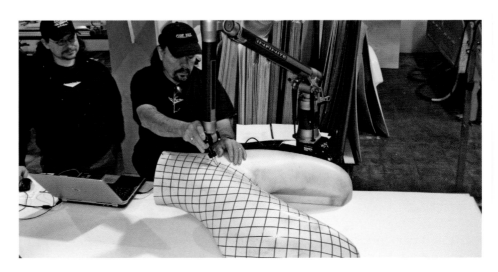

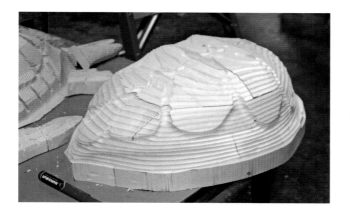

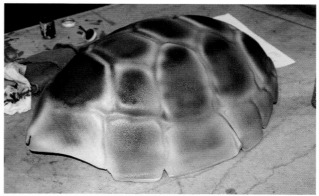

Above left and right: *A shot of a "plug" (left) and its part (right). This was made for the Discovery Channel.*

For a hammer form, you need to make the "positive" out of solid wood. To do this, you need to laminate a lot of smaller pieces together into one large block. We use a good wood glue and set the glued–and-clamped block aside to cure for a few hours.

Above left and right: *We use either Weldwood or Tightbond wood glue. There are many types on the market that will hold up well during the lamination process. We look for fast drying times, since we are always on some type of a deadline.*

Above left and right: *Even though we use a CNC router for our rough shaping, a few hours of hand sanding and scraping are still necessary to get the surface as smooth as possible. The wood we use for plugs is called Jelutong. It is a pattern wood and, oddly enough, is listed as both a hard wood and a soft wood. It comes from a gum tree in Malaysia.*

This is what is referred to as a "hammer form." It is little more than two pieces of wood clamped together, sandwiching a sheet of aluminum between them. The wood pieces are the shape of the part minus the flange distance. By hammering the outside edge of the metal over one of the pieces of wood, you get a nice flange without warping the flat sides of the metal.

MAKING WHEEL FLARES

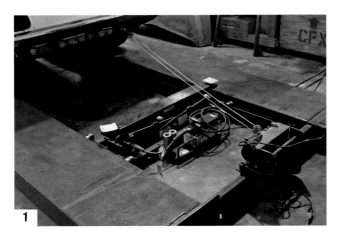
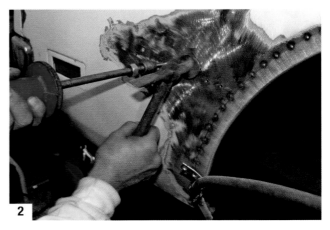

1 *To start the job of making steel wheel flares, I found it easier to put the car on a lift and bring it up to working height.* **2** *Before starting the flares, I need to get the body straight. There are a few light creases in the rear quarter panel that require a slide hammer and a hammer to repair.*

3 The flares have to be big enough to cover the MT tires that are going on the car, so the first part of the job is to mount the tires and cut out the body so they fit, and then build a wooden buck for the shape I want. **4** This buck is screwed to the body in a few places and will give the general form of the flare. It's easy to "mirror" the design for the flare on the other side of the car by making two identical sets of wood ribs and assembling them opposite of each other.

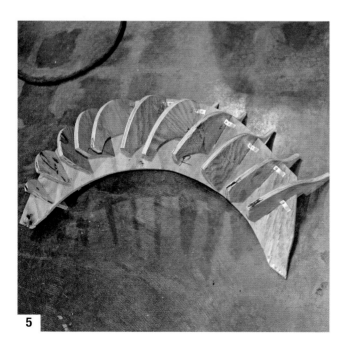

5 Bucks should be attached by only a few screws so they're easy to remove. This way you can get behind the metal to weld the individual plates as they are added.
6 In this close-up, you can see the angle brackets used to hold the rigs to the main radiused shape.

7 *A buck should be strong enough to allow you to make slight adjustments by pounding on the metal. Also note the bar used for the outside edge of the flare.* 8 *For the edge bar, I use ½-inch hot-rolled steel and contour it around a tire. Knowing it will spring out to a larger diameter, I use a smaller-diameter tire for the form.*

9 *Putting the bent bar on a workbench allows Brian to tweak it into shape before welding it to the car.* 10 *Now the bar is tack welded to the car body, and the buck is screwed into place.*

11 *We use a material called Vivex, a thin, clear plastic film, for most of our patterns. It allows us the ability to see where we are placing it and where we want to mark the edges for cuts.*

12 & 13 *Since the buck is a 3-D form and the plastic film used for the pattern is a flat shape, the film will not fit exactly. Cut it into sections (called gores) along the ribs for better fit. The buck will give you a visual reference of what the shape of each gore should look like as well as its placement.*

14 *As each of these sections is cut out of plastic and transferred to metal, the progression of the flares starts to take shape. Start in the top center of each flare and work forward and backward incrementally and form one side to the other, alternating as you cut each piece.*

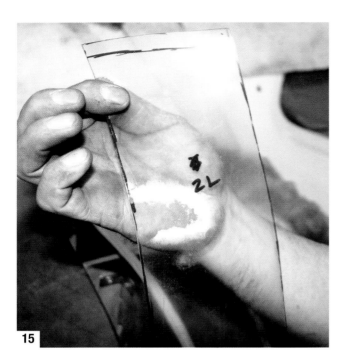

15 *Each gore should be marked and compared from one side's flare to the other. They should be roughly the same, or you have a problem. This is the time to resolve problems.* 16 *Never make all the patterns first, or they will not fit. Make the first patterns, one for each side, and then transfer the shapes to metal. Shape and tack weld the metal pieces into place on both sides.*

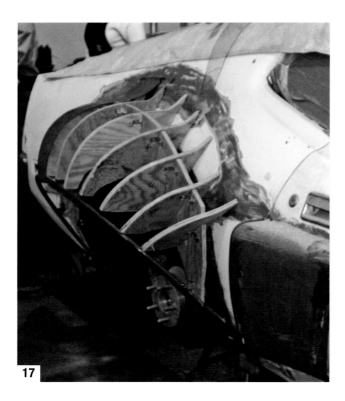

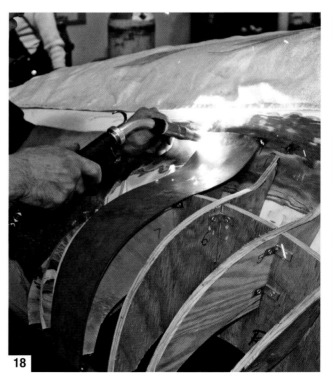

17 *A trial piece of metal is cut, formed, and placed on top of the buck to check the fit between the radiating ribs.* 18 *If you have a fitting problem, the first piece of metal can easily be removed, reshaped, and rewelded back on for the next pattern. If it was welded on and not tack welded, this would be a major event.*

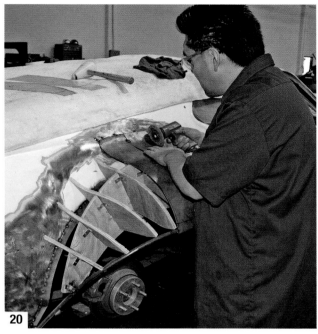

19 & 20 *Now make the next patterns and shape the metal for both sides, then tack weld those pieces into place. Repeat this process until both flares are complete.*

21 *A long-reach axial 4-inch cut-off tool is perfect for trimming and grinding. Ingersoll Rand makes an excellent heavy-duty tool for this process.* **22** *A Beverly shear is also handy for trimming each panel to the exact shape needed.*

23 *Each of the gores can be shaped off the car and then fine-tuned to perfect shape on the car, as we are doing here with a teardrop plastic hammer.* **24** *Never rush the job. Trim and shape as needed until you are 100 percent happy with the fit.*

25 *The wheelwell opening had to be modified, and this was a good time to do it. You will need to fit the tires and check for full travel.* **26 & 27** *As you can see, the process is fairly simple and repetitive. Just follow the steps: make the pattern in clear plastic, transfer that pattern to metal, cut the metal to the shape, form the shape to fit the buck, and tack weld the piece into place.*

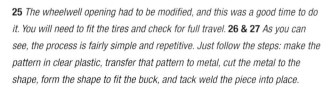

28

29

28, 29, 30, 31 & 32 *The finished job ready for final welding and bodywork. Each flare took about three days to fabricate to this point and will require about three more days of bodywork to complete. When finished, nothing will change the appearance of a car like a set of well-done steel flares.*

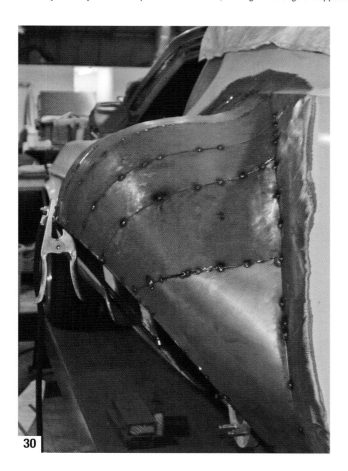

30

31

32

CHAPTER 4
METALWORKING TOOLS AND HOW TO USE THEM

ENGLISH WHEEL

Metal can be shaped in many different ways. Some shapes involve single bends. Others involve compound bends, in which the metal is bent in two different directions at the same time to create a convex or concave shape. Older automotive fenders like those found on any 1940 Ford or other '30 to '40s cars are prime examples of the deep compound curve.

Single bends can be performed with the edge of a workbench and a soft hammer, but if you want something more professional you may want to consider a means of making compound bends. Aside from the enhanced appearance of the smooth, flowing curve of a compound bend, the metal has more strength. This is due in part to the shape and also to the fact that the metal has been worked, which makes it work hardened. (Work hardening is a result of rolling or pounding metal or "working" it, compressing the molecules.)

However, it will require a reasonable outlay of cash or, as we like to say, an investment. I use the word "investment" because you will get a return on your money over time. Almost anyone who learns to use an English wheel can make good money making parts for others, or in savings by making parts for themselves.

Although the main tool you will need is an English wheel, a few supplementary tools will help rough in the shape before you start "wheeling" the metal. These additional tools include a sandbag, a few shaping hammers, a shrinker, and a stretcher.

In this chapter, I will cover the basic concepts involved in shaping metal through a few exercises that show the principles of the Eddie Paul English Wheel. We will start with some 12-inch squares of 3003 H14, a half-hard aluminum 0.063 inches thick. We'll anneal then wheel the metal into compound curves that can conform to whatever shape you may require. You can wheel both steel and aluminum, but we are using aluminum because it's easy to shape and can be annealed with a simple torch.

When fabricating body parts for cars, you will most likely be shaping steel. The process is the same, but for now aluminum lets you learn the process with less effort.

This is a section for the rear roof of my '40 ford, an example of what a few minutes of rolling will accomplish. A flat panel would not have worked, it had to be rolled.

When it comes to working with steel, you will find it takes more patience and more time.

Many people get the English wheel and the planishing hammer confused, and with good reason. The biggest difference is that the planishing hammer allows you to work smaller areas than the English wheel does. Also, the hammer is an air-powered tool, while the wheel is usually manual. We make and sell both of these tools, since they work hand in hand. A set of these tools turns the average hobby workshop into a professional metalworker's shop.

While a few companies offer English wheels in a variety of sizes, we have designed a tool that will fit the average small shop. It can be used to make parts like fender skirts, motorcycle fenders, or even motorcycle gas tanks. In fact, this tool will allow you to build almost any part of an entire car.

Features and Descriptions

The English wheel is normally constructed from a sturdy, "C"-shaped steel tube mandrel bent in a gentle curve, allowing space for your work piece, as the frame that holds the rolling (top) and anvil (bottom) wheels—somewhat like a big "C-Clamp." The depth of the "C" shape is called the "throat." Rolling wheels are flat. Anvil wheels come in different configurations, ranging from flat to crowned, to match the needs of the project at hand.

On an English wheel, the top wheel is always larger than the bottom wheel. and the bottom wheel is changeable for more or less curve.

This tool also uses solid steel rollers with bronze bearings for trouble-free extended running, or if you prefer you can get them with roller bearings at a slightly higher cost. Some of these tools are also designed to be portable and easy to set up, so they can be stored away after each use if needed; and some can even be converted from an English wheel into a planishing hammer within minutes utilizing the same frame.

An English wheel's use, like most metalworking tools, improves with operator experience. For this reason, I suggested that you practice on clean scrap metal before working on your project. I have had a metal supply company cut one-foot-square sheets for aluminum and have found the cost of the metal to be reasonable as a training cost.

The portable English wheel can be simply bolted to a workbench or mounted to a floor stand. The larger models stand on their own. but remember both must be bolted down to keep the wheel from walking away because of the pushing and pulling of the metal being shaped.

• Mount the wheel solidly

As with any tool that requires you to use both hands to move the metal, the tool must be mounted firmly for accuracy of control or the project will come out second rate.

• Keep the area neat and clean

Keeping the work area organized is more important than many think. In fact, many times during large movie projects we stop for "clean-up time." It is a chance to organize and relocate our tools while putting away the tools we don't need. In your case it will be a chance to just take a break and sweep up the area, wipe down the metal, and check on your progress. An organized area is much more conducive to a better-looking finished project.

• Keep the metal clean and deburred

The importance of clean metal cannot be stressed enough. It does more than make you look professional—it makes you act professional. Constantly wiping down the metal and wheels will eliminate the marring that occurs from dirt or other foreign objects getting between the wheel and the metal and leaving imperfections on the finished work. A single piece of dirt on one of the wheels can leave hundreds of indentations in a sheet of metal before it is discovered. So get in the habit of wiping down the wheels.

• Keep checking your progress against the form

Wheel lightly and check often so you don't overshape an area. Constantly check for fit on the metal and start getting in the habit of guessing how much wheeling will produce how much curve. Wheeling is about 80 percent skill and 20 percent muscle. Many consider it an art form, so a lot of the skill is in just getting a feel for how different metals react to variations in pressure and speed as well as numbers of passes.

• Less is better than more; use slow, light rolls

It is common to overwheel a part, rendering it scrap or, at best, difficult to repair. So remember: less is better than more when wheeling. Take light rolls, not heavy rolls. It seems at first that turning the bottom wheel tighter saves time, but, as you will learn, it only creases the metal, giving you more work blending out the crease. So take it easy and don't overdo it.

• Roll slowly to acquire the skill

Another skill that will come with time is the art of moving the metal back and forth, turning it just the right direction and just the right amount for the next run. The best way to gain this skill is to start out real slow and get the

This is an English wheel being used for an edge that needs to have a bit of a compound curve added to it. Just a little bit of rolling will add a lot of strength to an otherwise flat sheet of metal.

motion down, then increase your speed. This will take a lot of time but will eventually become second nature.

• *Mark the metal to show how you roll*

Don't be afraid to mark the metal to show where you want to roll. A high crown is marked on the front side of the sheet; reverse crowns are marked on the back side of the sheet. After you roll the sheet and check it, wipe it down to remove the marks then remark it as needed to help you keep track of the progress.

• *Change the roller to change the curve*

There is only one company that sells a template tool that will help you to figure out which bottom wheel you need to match a required form or curve. (It happens to be my company, EP Industries.) This template tool is to be used as a reference only and is not an exact template of the curve or shape. But the template will get you the general shape you need; the rest is up to you and your acquired skill.

• *Keep the rollers clean and polished*

As stated before, keeping the wheels clean is very important and can be as simple as wiping the wheels off from time to time. Or, worst case, you may need to chuck them up in a lathe or a drill press and spin them while holding a small piece of 2000-grit emery paper against the wheel. This will help keep them smooth and polished.

Things to Remember When Practicing

Practice will demonstrate how varying the pressure affects the tool's performance. This is a personal preference, but we suggest you start out at low speeds and only go faster as your skill level increases. Lower pressure is good for smoothing, while high pressure is good mainly for rough shaping because it leaves larger wheeling marks.

Practice with aluminum, since it is the easiest metal to work with and will teach you the basics with less effort than steel. Later, you can apply your newly learned skill to autobody sheet metal. The best alloy of aluminum for practicing is 3003 H14 half-hard aluminum 0.063 inches thick. Other alloys can be used, but this grade is the best for welding and working.

Keep the metal dry, or it will attract dust and dirt. With every move of the English wheel, the metal is stretched a little bit over the lower wheel crown, creating the curved shape. Therefore, the more curve there is in the lower wheel's crown, the more curve you'll add to the

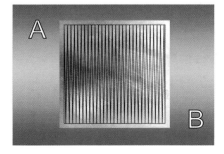

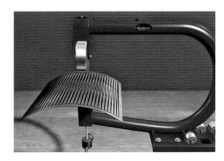

To roll a single bend, simply follow a zigzag pattern across the sheet of metal, moving from a starting point and going to the end point in one direction as shown. The curve will vary in proportion to the number of passes you make. The more passes, the tighter the curve.

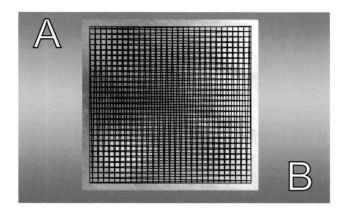

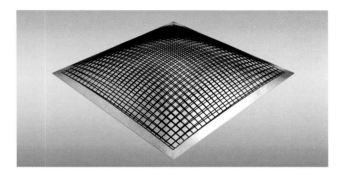

If you run the wheel in a second pattern perpendicular to the first pattern, you'll get a compound curve. As with the simple bend, the curve depends on how many passes you make.

metal being shaped. The flatter the crown of the lower wheel, the less curve you'll add.

Try not to overwork an area. Removing a high crown is not as easy as putting one in. Work slowly and double-check your progress often, especially if you are making a body panel.

Basic Tracking

Now, let's start shaping the metal. We'll begin with a basic example. The ideal tracking path is a series of tight zig-zags, spaced about ⅜ of an inch apart. Start at one corner, call it point A, and slowly work the metal back and forth until you reach the end at point B; now try going in reverse of the pattern you made by tracking the metal back to point A using the same zigzag pattern. Cross wheeling the metal is exactly the same, but with the piece turned 90 degrees.

This is the essence of shaping with the English wheel. You should refrain from rotating the metal being shaped while it is between the wheels, as this will mark up your work.

If all the tracks in a tracking pattern end up on the same line, the effect will be very noticeable. Instead of a smooth blend from one shape to another, there will be an abrupt transition. This can be avoided by using the Staggered Stop method. This technique uses two different stop points for the track lines, blending the newly raised metal and the old shape much more smoothly. On some larger projects, you may have three or more stop points in order to create the desired shape.

PLANISHING HAMMER

The planishing hammer uses a pneumatic riveting hammer, not an air hammer. This is a much more expensive tool, but it has a longer life and a much faster cycle. A common air hammer will not work as well and should not be used. The planishing hammer can be used for raising a section in the center of a piece of metal and is commonly used in conjunction with the English wheel. This is why some manufacturers sell them as a set. Generally speaking, I use the wheel for long, gentle domes, such as a car roof or the top of a gas tank, and the hammer for the sides of a gas tank (higher crown area) or a hood scoop on a car.

The planishing hammer is very loud, vibrates a lot, and often comes apart at the bolts, so you will need to constantly re-tighten bolts and wear hearing protection as well as goggles. Also keep in mind that in the process of planishing an area, the metal is being displaced and becomes slightly thinner and much harder in that area. So the area may need to be annealed from time to time until your work is complete.

ANNEALING THE METAL

You will notice that the more the metal is shaped, the harder it becomes. So be prepared either to spend more time on areas that have been worked or anneal the metal between sessions.

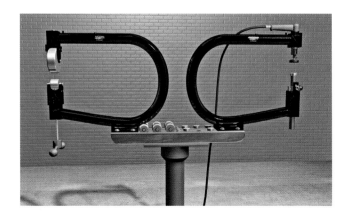

The planishing hammer/English wheel combination set E.P. Industries makes and sells. It allows the fabricator to form just about any shape in metal with one set of tools.

A planishing hammer. The planishing hammer smoothes out rough pounded metal, expanding it slightly depending on the radius of the bottom anvils.

By changing the bottom anvils, or dies, you can change the radius of the compound curve you are hammering into the metal. When purchasing a planishing hammer, you should also get a set of dies. Most planishing hammers come with dies in three different radiuses.

It is a good idea to anneal the aluminum, because it will allow you to form the metal easily. As you work the metal, you will find it gets stronger and harder. This is referred to as "work hardening" and is the result of the metal's molecules being compressed together by repeatedly rolling the metal between the top wheel and the bottom wheel. The compressed molecules result in a denser metal. Annealing the metal will allow the molecules to move farther apart, softening the metal.

All you're really doing is heating up the metal. As everyone knows, heating a material causes it to expand. This is because the material's molecules move around more as they get hotter. When they move around, they require more space, thus the material expands.

Before annealing the metal, put a slight roll to the metal, or it will wind up buckling and becoming harder to work. By putting a slight roll to the sheet, the buckling can be controlled and minimized.

The process of annealing is simple enough and only takes a few minutes. The only tool required is an acetylene torch with a rosebud head. First, adjust the torch to a "dirty" or pure acetylene condition and "paint" the area to be annealed with soot.

Only paint the area to be annealed; don't get carried away and paint the entire sheet. This soot-painted area will act as a temperature sensor to help you determine how much heat is being applied to the area you are annealing. The amount of heat needed to burn off the soot is the same as the amount of heat needed to anneal the aluminum. If too much heat is applied during the annealing process, you will melt the aluminum. On the other hand, if you do not apply enough heat to burn off the soot, the area will not have been heated enough to anneal it.

After you've painted the area to be annealed, adjust the torch to a neutral flame and burn off the soot. Move the torch constantly, so you do not apply too much heat to any one area and melt the metal. A slow, circular path will control the heat and burn off the soot. When you are done heating the material, it can be quenched in water. (The exact parameters of this procedure will be determined by the makeup of the alloy, so check with the metal supply company on the particular metal you are using to find the proper method of annealing.) The area should now be annealed. Simply bending the edge of the sheet should confirm this.

OTHER TOOLS

We will also be using a few other tools in conjunction with the English wheel. You do not have to have the actual tools, and you can improvise if you want, but they will enhance the use of the wheel. Panel beater's sandbag, dollies, tear-drop mallets (these come in different sizes), round mallets (these come in different sizes), manual shrinker, and manual stretcher.

Using the sandbag is a good way to quickly rough in the desired shape of your project. It is used in conjunction with the mallets. Hold the sandbag behind the metal and strike the front of the metal with the mallet. When the mallet strikes the metal, it stretches the metal, causing it to bend. The sandbag itself absorbs the blow from the mallet while still allowing the metal to bend.

In order to create the desired shape, you must use force to shape the metal. The real trick is learning how much force must be applied. Several well-placed taps are much better than one large, forceful blow. Also note that you need to "fluff" the sandbag often in order to keep the sand from pushing out to the sides, reducing the effectiveness of the bag.

Another useful metal-shaping technique involves the use of manual shrinkers and stretchers. Using a shrinker on the edge of the metal creates a higher crown.

The planishing hammer is another tool that works hand in hand with the English wheel. By changing the crown of the lower anvil and the pressure between the two hammer heads, it is possible to create a great variety of curved shapes.

The "generic" planishing hammer is not unlike the English wheel in its overall look until you look closely at the area that meets the metal and where the English wheel has rollers that press the metal as the metal is passed thru the two rollers. The planishing hammer has a hammer on the top and an anvil on the bottom. These are a pneumatic version of the body man with his hammer and dolly, only much faster and much more controllable, to a point. That point being that you cannot do hammer-off shaping. Only hammer-on shaping.

Hammer-off shaping is where you place the dolly under the metal, slightly to the side of the hammer blow at the low spot in the dent, and tap with the hammer near the dolly but not directly above the dolly (at a high spot) moving two areas at one time. The high spot should be down and the low spot up. The planishing hammer only hits directly above the anvil (dolly) and, in doing so, will expand the metal in this area, raising a crown upward that follows the curve of the lower anvil. That is why the lower anvil is so easy to change—because it commonly *is* changed when moving from working one area of the metal to another.

SHRINKING AND STRETCHING

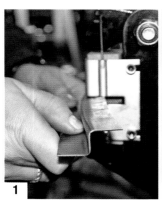

1 & 2 *For shrinking and stretching along an edge, we use a two-sided pneumatic unit. One side shrinks metal; the other stretches it. Here we shrink an edge to begin forming a radius.*

3 *Now the tool is rotated so the other end is accessed to stretch the outer edge.*

4 & 5 *By simply stretching one side and shrinking the other, we can make a round section for repairing door-frame corners. This process only takes minutes to perform with the right tool.*

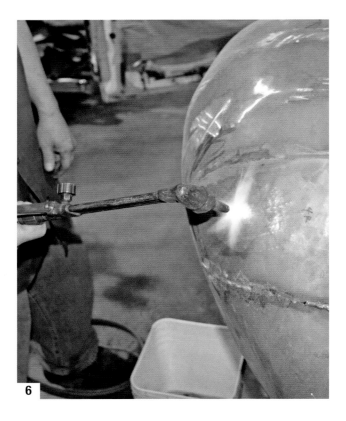

6 If you need to shrink metal in the center of a panel, the best method is to use a torch. Apply heat to an area about the size of a silver dollar. Get the metal red hot, making it expand and raising it up.

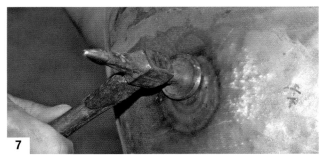

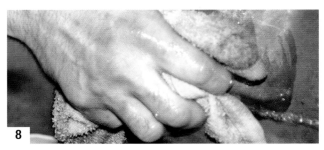

7 While the metal is still hot, place a dolly on one side of the heated metal and tap the other side slightly to compress the metal into itself, thus shrinking it.
8 Now the controversy: Do you cool the worked area with water or let it cool by itself? It depends on if you are planning to do more work in the area. Cooling the spot will harden the metal, making it hard to rework if needed. But if you leave it hot, you will wind up burning yourself eventually by inadvertently rubbing up against it. I usually opt for the cooling method. You can always reheat and anneal the metal for later work if needed.

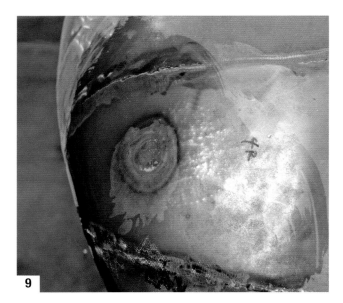

9 This is what an area looks like after shrinking. This process can be repeated in as many spots as needed. **10 & 11** Other methods require shrinking hammers and dollies, and they work fairly well on areas that only require a bit of light shrinking. But if you need extensive shrinking, you will need a torch.

METALWORKING TOOLS

Left: The Morgan Nokker has been the bodyman's slide hammer of choice for several decades. Morgan Manufacturing offers several attachments and slide weights for handling a variety of repairs. **Right:** You'll find about a half-dozen plasma-cutting machines in my shop, ranging from portable 110-volt units to larger 220-volt machines capable of slicing half-inch steel or thicker. This is the Powermaxx 1000 from Hypertherm, one of the best on the market.

Left: Panel alignment is a fairly simple task that requires some basic tools. You'll need wrenches for loosening and tightening bolts, of course. I also recommend a panel-alignment gauge and these plastic prying tools from the Eastwood Company. They will allow you to fine-tune your panel alignment without damaging your paint. **Right:** We use this old horizontal band saw to make precision cuts in thick chunks of metal. A good horizontal band saw will have an automatic sprayer for cutting lubricant.

Left: *The presence of computer numeric controlled (CNC) machines is a sign of the times. This is the latest CNC router from K2 CNC. We have two of these to keep up with our hectic deadline schedule. CNC routers are great for turning sheets of plywood into metal-forming bucks for shapes with compound curves.* **Right:** *The pneumatic dual-action (DA) sander is a must for paint preparation as well as body and metal finishing. These two sanders are made by Hutchins Mfg. (the blue model 4500 has a finer orbital pattern than the red model 3500), and the sander on the car is made by National Detroit. I recommend you visit these companies' websites for a look at their full line of products.*

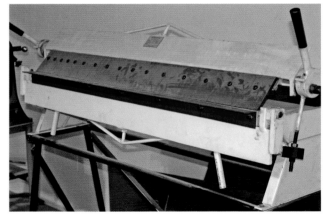

Above: *The infrared heat lamp can be a painter's best friend when weather takes a turn for the worst. This basic model, available from the Eastwood Company, is compact and affordable. Higher-end models will have multiple elements and heat control.* **Right:** *A metal brake is standard equipment in any metal fabricator's shop. Small (less than 4 feet wide) imported models like this are good for jobs requiring 18-gauge sheet and thinner. The fingers are removable to accommodate shapes.*

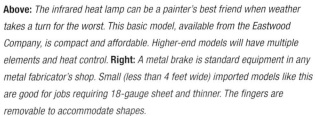

Left: *A small 3-pound slide hammer is the tried-and-true way to work out a crease in the side of a car. Pros: cheap, poratable, no power required. Cons: you must drill and fill ⅛-inch holes for the slide attachment.* **Right:** *A product of modern technology is the laser line level. It is available at most hardware stores and, if used properly, can help with numerous car-related projects.*

Left: *A variety of work requires a multitude of tools. Saws come in all sizes and configurations; this one is a pneumatic model made by Chicago Pneumatic. The blade is small enough to allow good maneuverability, and the tool itself can be held with one hand.* **Right:** *When working on cars, drilling or fastener access is often limited, requiring either manual tools or a handy right-angle drill-driver. DeWalt is one of the only tool manufacturers that offers an 18-volt, nickel-cadmium, battery-powered, right-angle drill/driver.*

Left: The Scotchman tube-notching machine uses an abrasive belt with different roller sizes to match the notching requirements of a variety of tube diameters. It can also be used as a regular belt sander, as shown here. **Right:** A dry-cut saw, such as this 14-inch model, makes easy work of cutting metal stock. **Below left:** The tool that you see here is called the slide sledge. When the components are assembled, the slide sledge offers a more accurate and controlled hammering of body parts. **Below right:** The Bosch Tools Model 1521 is a 16-gauge metal-cutting shear that offers both power and speed. Minimum curve-cutting radius is 1.25 inches.

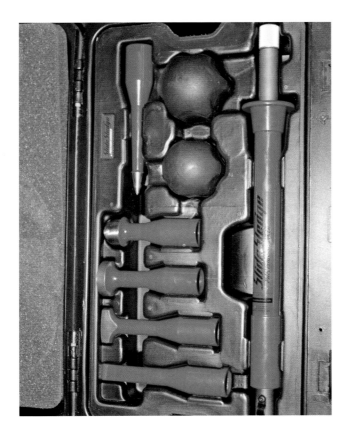

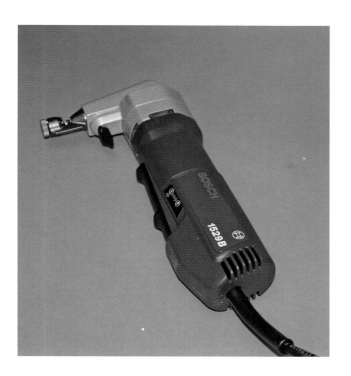

Left: *Bosch manufactures electric metal-cutting nibblers in four models to handle sheet metal from 10- to 18-gauge.* **Right:** *The CPA835 from Chicago Pneumatic is an air-powered nibbler rated for 16-gauge (.060 inch) steel and aluminum up to 10-gauge (.102 inch).*

Left: *The Bosch Model 1873-8 is a large 7-inch-diameter grinder with a 15 amp/8,500 rpm motor and tool-free adjustable guard for making quick changes in working position.* **Right:** *These are Bosch cordless saws (shown without battery) from their 18-volt nickel-cadmium series kit. The saber and reciprocating saws have tool-free blade-changing features and use the Bosch Blue Core battery technology. The Bosch 1662B adds a 6.5-inch cordless circular saw to the lineup. Bosch also makes 24-volt versions of these tools.*

Left: *This multiple bay charger from Bosch will charge both their 18-volt and 24-volt batteries.* **Right:** *Many old foot-operated shears, such as this, are still in use. This one has endured over 20 years of cutting and abuse and is still going strong.*

Left: *The leverage of pedestal-mounted hand shears makes effortless work out of cutting sheet metal.* **Right:** *Left-, right-, and straight-cutting aviation snips from manufacturers like Bessey and Irwin are still used by metalworkers for convenient trimming and cutting of sheet metal up to 20-gauge.*

Above: *Here I'm using a body hammer to straighten the edge of a freshly cut piece of metal.* **Above right:** *Even with all the different types of cutters and shears at our disposal, sometimes rough edges need to be fine-tuned with a small right-angle die grinder like this Chicago Pneumatic model that features a unique 110-degree grinding head angle.*

Left: *Another old favorite at Customs by Eddie Paul that has been around for decades is this finger brake. With a well-thought-out order of bends, this tool can create myriad shapes.* **Right:** *Magnetic torpedo levels and magnetic squares are invaluable for metal fabrication. Bessey Tools offers a high-quality line of magnetic tools for all your metalworking needs. They're especially handy when you're working with several pieces or angles and need an extra pair of hands.*

Above: The Eastwood tubing roller, manufactured in the United States by E.P. Industries, is a cost-effective tool for manual bending of metal tube up to 2 inches in diameter and wall thicknesses up to .083 inch.

Left: Also a must for metalworking is a dedicated table for fabricating and welding. We made this welding table with a ½-inch plate of mill-flat steel placed on a heavy-duty table with casters.

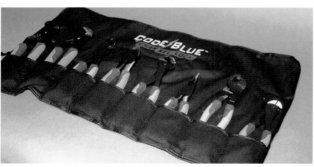

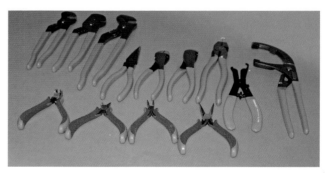

Above and above right: The familiar blue-handled pliers from Channellock now includes the Code Blue line with new and comfortable, blue-and-red ergonomic grips. The regular line of Channellock hand tools is as diverse as ever, with specialty pliers for oil filters, electrical work, and general mechanics.

Right: Every toolbox should have a pair of needle-nose pliers, side cutters, high-leverage side cutters, and lineman's pliers.

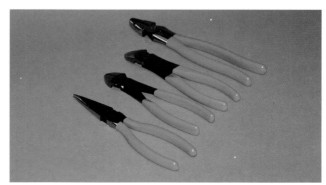

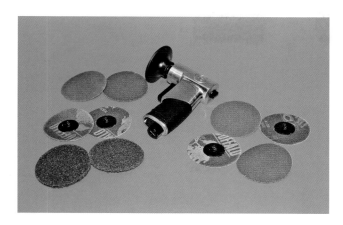

Above left and right and below left and right: *Chicago Pneumatic offers a full range of sanders and grinders in sizes ranging from compact right-angle tools pictured here to heavy-duty industrial units.*

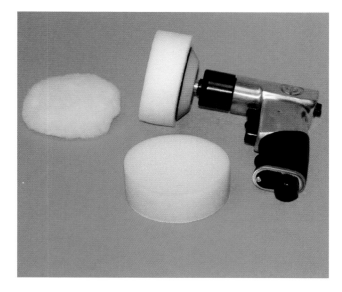

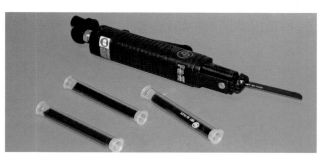

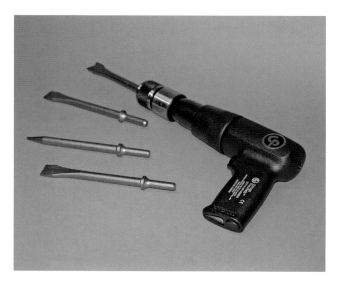

Above: *An air chisel can be used for so much more than chiseling. Armed with a basic collection of bits, an air chisel can cut, scrape, punch holes, hammer, and, of course, chisel.* **Right:** *The Chicago Pneumatic reciprocating saw is air operated, is rated to deliver 10,000 strokes per minute, and has an adjustable guide to set depth of cut—an excellent tool for cutting metal as well as plastics.*

All of your threading needs can be met with a top-quality kit like this one from Irwin, makers of the famous Vise Grip.

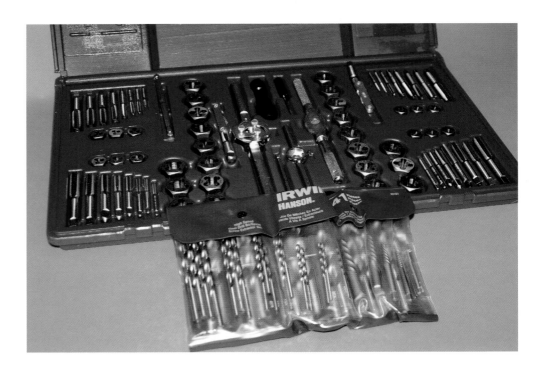

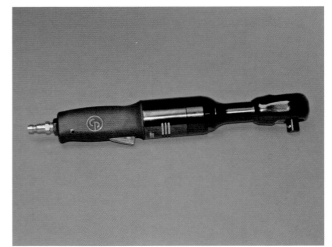

Above: *Chicago Pneumatic's CPA7300 is a tiny but powerful ¼-inch drill. Its 2,500 rpm motor is reversible and quiet.*

Above right: *Both autobody workers and mechanics rely heavily on a good air ratchet. Companies like Ingersoll Rand and Chicago Pneumatic offer several sizes and models to cover various applications.*

Right: *One necessary item in the metal fabricator's box of goodies is the air cut-off tool.*

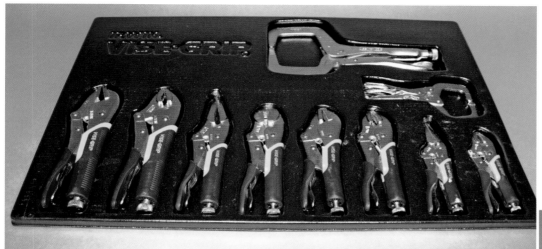

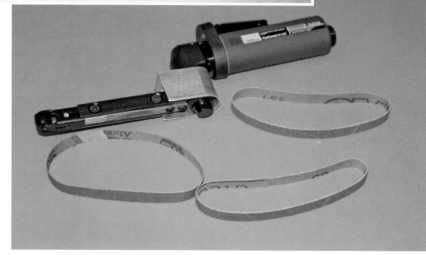

Above: *You can judge a metal fabricator's capabilities by the number of clamps that are in the toolbox. A top-quality starter set is this one from Irwin, makers of the Vise Grip locking pliers.*

Right: *This ultra-handy air-powered belt sander from Ingersoll Rand features a small, versatile 10mm belt operated by a 15,000-rpm air motor.*

Below: *CP enters the palm sander market with this 6-inch random orbital model in both standard and vacuum configuration.*

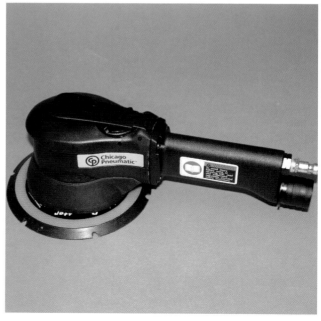

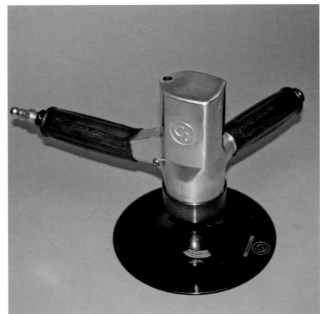

Left: *CP also offers a standard dual-action sander with the traditional handle and six-inch pad.*

Right: *Heavy-duty polishers and high-speed sanders come in these two styles from Ingersoll Rand and Chicago Pneumatic. These air-powered tools are favored by many metalworkers and painters due to their light weight and cool operation over extended run times.*

Below: *In addition to their Vise Grip line of locking pliers, Irwin Tools offers adjustable pliers, wrenches, needle-nose pliers, and cutters with a distinctive Irwin blue-and-yellow grip.*

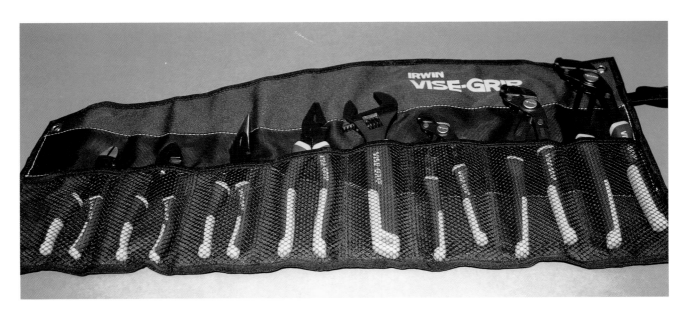

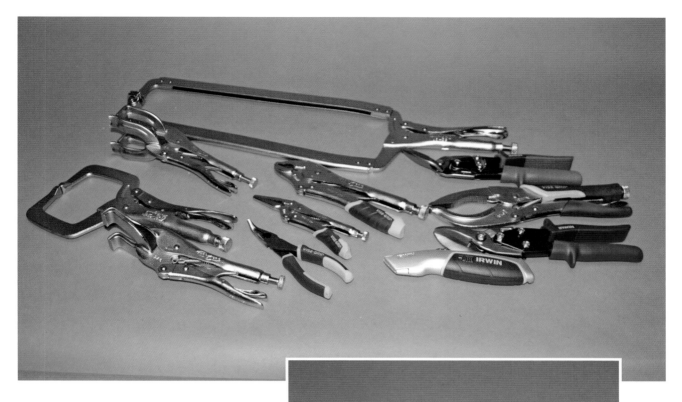

Above: *Here's a look at an assortment of Irwin's locking Vise Grips, adjustable pliers, and high-leverage metal-cutting snips.* **Inset:** *Identification of cutting direction is standard, green for right, red for left, and yellow (not pictured) for straight. The cutter in the center is an Irwin utility knife with a quick-change blade feature. Not exactly a metalworker's tool, but a handy one nonetheless.*

Above: *Toting your hand tools around the shop or garage is a pleasure with Irwin's tool caddy.* **Right:** *A set of step drills and drill bits ranging from ⅟₁₆-inch to ½-inch will cover just about every drilling requirement.*

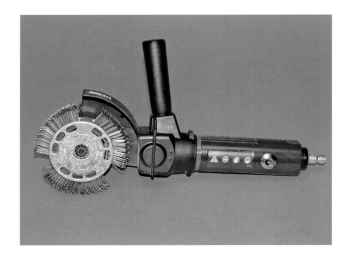

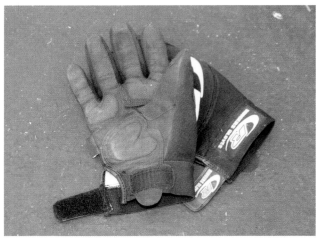

Left: *The MBX Power Tool is a pneumatic wonder that quickly removes rust, corrosion, paint, undercoating, and more from many surfaces. It operates at 3,500 rpm and comes with several different brushes/belts for specific applications.* **Right:** *Workers in my shop wear Ringer Gloves for safety and to maintain a good grip on their tools.*

Left: *The Monster Wheel manufactured by E.P. Industries is a full-size heavy-duty English wheel for the master metal fabricator. The Monster Wheel has a super-sturdy frame and a variety of optional attachments and dies.* **Inset:** *A drawer full of different kinds of snips, even old rusty ones, can come in handy when working with sheet metal.*

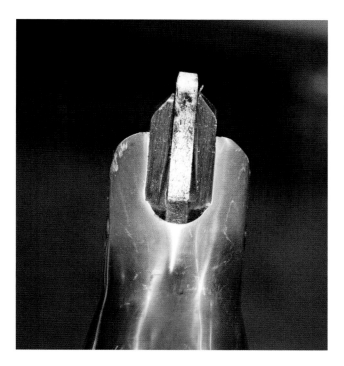

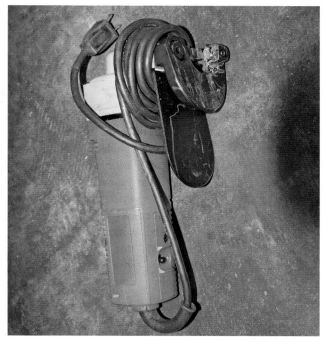

Left: *This is a close-up of the three-bladed shear that removes about ⅜ inch of metal from the cut but does not warp the metal as it cuts.* **Right:** *Milwaukee Electric Tools manufactures an extensive product line including tools for the metal fabricator. This is a Milwaukee #6805 16-gauge shear that can cut fast and handle tough jobs, even stainless steel up to 18-gauge. Milwaukee model #6815 is a similar tool rated for thicker 14-gauge, while the Milwaukee #6852-20 is a three-blade type shear rated for 18-gauge.*

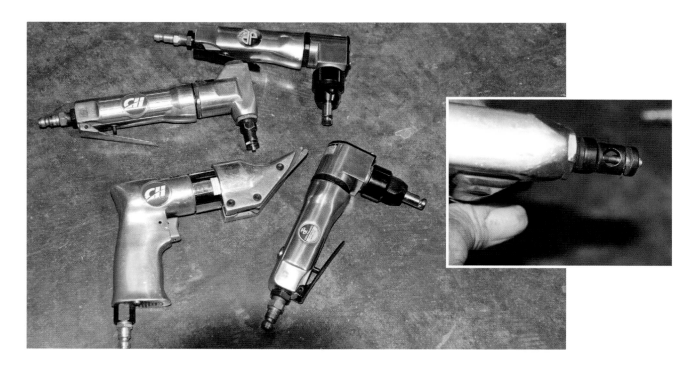

Left: *Discount tool outlets and your local hardware store can be great sources for tools.* **Inset:** *Be sure to maintain your pneumatic tools properly or they will end up looking like this. A periodic drop of lubricating oil will increase the life of air tools dramatically.*

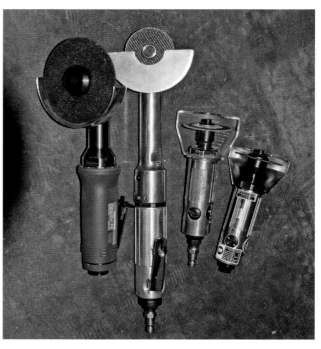

Left: *To avoid frustrating yourself during a project, I recommend organizing your tools by type and then neatly arranging them in a drawer. While this may be where all the hammers are kept, it will be difficult to reach the ones at the bottom. Having a lot of tools requires some attention to details, or using them will not be fun or productive.* **Right:** *This assortment of cut-off tools shows two basic styles: standard inline and extended reach.*

Left: *Just about everybody has a Sears store or an OSH Center in their town where Craftsman tools can be found. If not, you can purchase items like this Craftsman air compressor on the Craftsman website.* **Right:** *The Craftsman website and catalog offer the full range of hand tools and power equipment, including this industrial-quality floor-standing drill press with all the bells and whistles.*

Above: *A sturdy, centrally located worktable with nearby power outlets is essential for items like chop saws.*
Inset: *When working with long lengths of material, these cheap roller stands are quite handy.*

Left: *A hydraulic bending machine, such as this model from Mittler Brothers, will allow you to make precision bends in heavy-wall steel tube for projects like frames and roll bars.* **Right:** *A big, heavy shop vise is one of the essential fabricator's tools. It must be sturdy and properly mounted so that you can clamp down on any object, regardless of size or weight. This Columbian is over 45 years old, salvaged from a downtown Los Angeles Ford dealer back in the 1960s. It still serves us well today.*

Lifting and support options include these floor jacks and stands from the Bend Pak/Ranger company. They make a full range of lifts, from portable to large floor-mounted units, as well as a variety of other specialty equipment. Once you've raised the object, support it with jack stands.

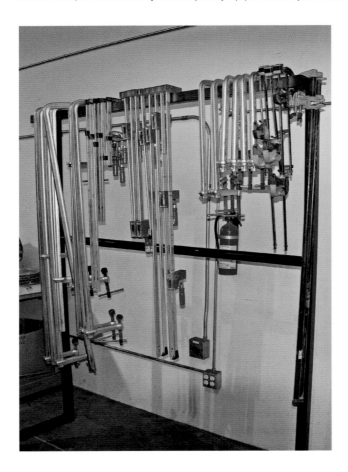

Above: Another item from the Eastwood catalog is this vintage-style read roller. We've modified ours with the addition of a Grant steering wheel to facilitate the rolling process.

Left: Once you have some basic fabricating skills, you can begin to build your own racks for your tools. This is where we keep our assortment of Bessey Clamps for welding, woodwork, and other projects. Bessey products are top of the line.

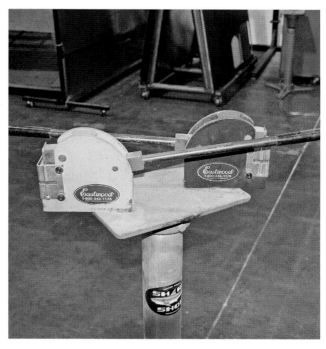

Left: *Although a little dusty, this sheet metal roller sees regular use at Customs by Eddie Paul.* **Right:** *Conveniently mounting the Eastwood manual shrinker and stretcher on a grinder pedestal keeps them ready for use.*

Left: *This E.P. Industries English wheel is the midsized unit that is a perfect tool for fabricating shops and serious home enthusiasts. It comes with three anvil sizes and a helpful how-to tape to teach you how to use it.* **Right:** *The E.P. Industries planishing hammer is another must-have for metal fabricators. Whether you're making custom chopper gas tanks or patch panels for your car, this is the professional tool to use.*

Left: *A sheet metal notcher like this will make short work out of cutting corners and angles.* **Right:** *Band saw blades can become a costly expense of you do a lot of cutting. In the long run, it is much more cost-efficient to purchase bulk blade stock and use a blade welder to manufacture your own.*

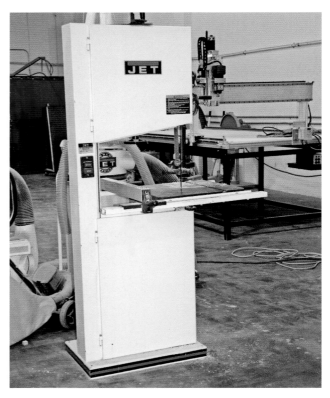

Left: *Some of the best tools are the old ones. This Do-All vertical band saw has been in service for many years.* **Right:** *This newer Jet band saw is fitted with a special blade for cutting plastic and wood. A shop must have dedicated equipment to avoid time-consuming blade changes during the day.*

Above: *The Bend Pak MD-6XP Scissor Lift has a 6,000-pound lifting capacity, which makes it perfect for lifting cars and light-duty trucks. It is portable and operates on 110/220-volt power.*

Left: *This worker is using the original Vise Grip Large Jaw Locking Plier to hold his work in place while TIG welding. As I've said many times before, a good fabricator will have many clamps and locking pliers in his toolbox. You can't have enough.*

DIMENSION DEFINITION PART GEO   Z END   0.0000
CONRAD   0.0000
TOOL OFFSET   CENTER

CHAPTER 5
BIG EQUIPMENT FOR METALWORK: CNC LATHES AND MILLING MACHINES

By big, I mean big! We use a lot of larger machines, not only for making parts but for making tools to make parts. So, major tools are a part of metalworking, and you should at least have a working knowledge of them if not an affinity for speaking "G-code" to impress people at the next party you attend. Cartesian coordinates will become a way of life, and you will find yourself giving directions to your friends using incremental and absolute distances. In other words, you will be come an absolute Mech-O-Nerd, and before you ask, no they do not give out belts for fourth-degree Mech-O-Nerds.

COMPUTER NUMERIC CONTROL (CNC)

Today's high-tech mills and lathes are run by computer numeric control (CNC), which means the machine is directed into the raw metal by codes generated by a bunch of ones and zeros precariously colonized into an assemblage of order through a program sometimes called "machine language" or just "program." There are a few different ways to write a program. You can memorize the language or code called "G" and "M" codes that tell the machine to do certain things at certain times, like drill a hole and use a number of pecks during the drilling cycle. Or you can program the machine to go from one point to another point while cutting at a set speed.

They also have what is called "conversational language," which I prefer. It is very simple and is just an interface for those of us who do not mind getting our feet wet but are not ready to take the plunge in G and M codes.

Conversational language is kind of like talking to the machine and answering a few questions that it asks you. For example, it might ask me "What would you like to do today, Eddie?" A list of events are then shown on the screen, like DRILL, MILL, BORE, or POCKET. So, I hit the button under the DRILL command, at which point the machine asks what size hole I require, and how deep will I need the hole, how many pecks would I like, at what

By having my own in-house machine shop, I am no longer at the mercy of other companies for parts or tools, since I can build almost anything I need. Most of these tools can be picked up at machinery auctions very inexpensively.

speed do I want the chuck to turn, as well as a host of other questions. Like, "Oh yeah Ed, where do you want the hole?"

After it's done asking questions, you grit your teeth, say a fast prayer, and hit the start button. If your deity of choice is listening, a hole magically appears in the metal where you wanted it to appear, plus or minus a few thousandths of an inch, depending on the quality of the CNC mill.

Now you just add more events. You can make really complex parts by combining drilling events, milling events, and cutting events.

So, with CNCs, almost anything you can think of can be machined with extreme accuracy and repeated numerous times, exactly as programmed.

As for the terminology of CNCs, I could go on for the rest of the book and still not tell you everything, but the basics are as follows.

Incremental measurements are measurements from the last position you were at to the next position you would like the mill to go.

Absolute measurements are the measurements from one predetermined location to all the locations you would like to go with the cutter.

For example, if you go from your home to the store and then to the gas station, you can have many incremental measurements but only one absolute mesurement. You could go one mile from your home to the store. That's one mile absolute and one mile incremental. If you then go to the gas station, which is a mile further down the road, you have gone one more mile incremental (from the store to the gas station) but two miles absolute (from your home to the gas station). But that station is closed, so you go to the one two miles further down the road. That's two miles incremental from the first gas station, or three miles incremental from the store, or four miles absolute from your house. Again, absolute is from the starting point, and you can only have one starting point.

CARTESIAN COORDINATES

The mills and lathes all work on three dimensions in space—right and left, up and down, forward and back—and with these three dimensions you can go anywhere. But you need a name for the directions, not just left and a little ways up, even though that would kind of work. So, we have something called Cartesian coordinates to represent directions from the operator's point of view while looking at the table.

Remember our example about going to the store and then to the gas station? And, we started from home, so everything was expressed as a distance from home? In the Cartesian system, home is called the origin. Everything is measured from the origin.

In two-space, there are two axis, the X axis (left and right), and the Y axis (up and down). These two axis intersect at the origin, which is denoted as (0, 0). In three-space, there are three axis, the X axis (left and right), the Y axis (forward and back) and the Z axis (up and down). Note that the Y axis has changed orientation in three-space. These three axis intersect at the origin, which is denoted as (0, 0, 0). Coordinates along the X axis are expressed as positive numbers to the right of the origin, and as negative numbers to the left. Coordinates along the Y axis are expressed as positive numbers as they come toward you from the origin, and as negative numbers as they move back. Coordinates along the Z axis are expressed as positive numbers above the origin, and as negative numbers below.

Now this is just a rough description of the very basics of the system. If you get the right machine and have a good instructor, you can be up and running the same day

This is one of my CNC mills that will run three-axis programs using conversational programming language. This makes it very easy to program and run.

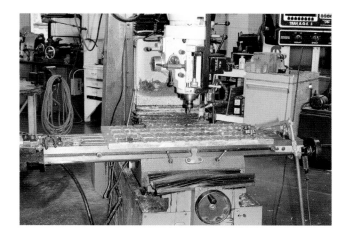

On regular as well as CNC mills, the bed is what travels on the X and Y axes, and the quill travels in the positive and negative Z axis. The longer the X, Y, and Z travel, the more the cost of the mill, but the bigger the jobs you can run.

the machine is installed. Or it could take you about a week to be programming and running parts. In any case, the example I often use is that the machine will more than pay for itself if you are just the least bit productive with it. So once again, do not think of a CNC as a cost but as a short-term investment in your future.

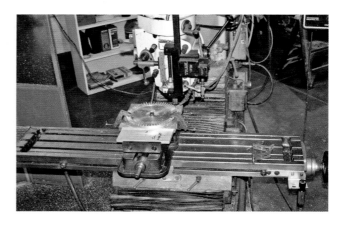

This is a heavy-duty Fadal CNC vertical machining center. Its enclosed cabinet lets you continually spray the work with coolant without getting it all over the shop, so you can cut deeper and faster.

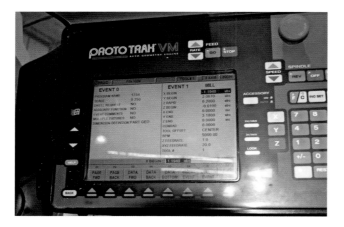

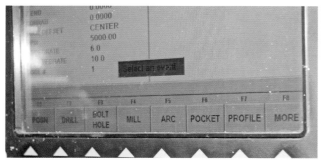

This is the control panel on the Proto Trak VM, where you program the machine to do what you want it to do. All you have to do is answer a few simple questions, and the machine translates into G-code for the machine to follow. This is called conversational machine language.

Cartesian coordinates were invented by Rene Descartes (born in 1596) as a method of locating an object in three-dimensional space. We use them in our machine shop as a way to tell the machine where to move when cutting. All CNC machines are programmed to understand them, so there is no escaping the fact that you need to learn the system. In fact, you will find it is not all that complicated. The Cartesian coordinates are shown, with the red arrows indicating the positive Z (up) and negative Z (down) axis (x, y, ±Z) and the blue arrows indicating the positive X (right) and negative X (left) axis (±X, y, z). The yellow arrows indicate the Y axis, with the positive Y axis pointing away from you and the negative Y axis pointing toward you (x, ±Y, z). The point at which all the axes meet is called the origin and is represented numerically as (0, 0, 0)

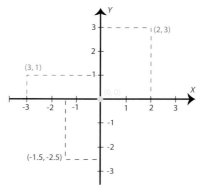

Cartesian coordinate system with the circle of radius 2 centered at the origin marked in red. The equation of the circle is $x^2 + y^2 = 4$.
www.wikipedia.org

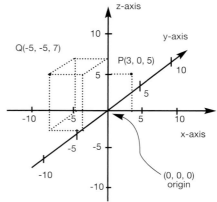

Three dimensional Cartesian coordinate system with y-axis pointing away from the observer.

CHAPTER 6
CUT, BEND, AND SHAPE

CUTTING TOOLS

Cutting tools can include a host of items, from simple snips to plasma cutters and everything in between. In fact, many shops are using the wrong cutter for the job, such as OCC using a small abrasive cutter for just about everything even though there are better tools for the job than that one. Some people feel comfortable with certain tools, so they keep using them instead of using tools that would result in a more professional-looking finished product.

Cutting by shearing is mainly used on thinner sheet metal. There are a number of tools that perform this job quite nicely by using a shearing action between two or three blades. Another type of cutting tool is called a "nibbler." They are either electric or pneumatic and use a nibbling action to cut out small half-moon sections from the sheet of metal.

Cutting with abrasives is standard practice on most of the TV shows, but it's not the best way to cut sheet metal because it produces a lot of sparks that can get in your eyes and a lot of fine, abrasive dust particles that float around the shop for a while, then settle and ruin engines or other precision equipment. You'll also burn through abrasive discs at an alarming rate, which gets costly. In some cases, though, an abrasive blade is just the tool for the job, since it will cut rusty sheet metal as well as it cuts through a frame. It is a tool that cuts anything you can reach, it just isn't very precise. It is a lot like using a crescent wrench in lieu of the right open- or box-end wrench. It works, but you could use a better tool. Having said that, I have about a dozen abrasive cutting tools in the shop. I often grab one of them to tear down old cars, since I can cut the heads off rusty bolts or cut off a damaged fender to give me better access to the mounting bolts, saving me a lot of time.

Cutting with heat involves cutting torches and plasma cutters. Both these types of cutters basically use heat to bring the metal to a molten state, then the oxygen increases the temperature and blows out the molten metal. The

For cutting metal to a rough shape or just getting a straight edge, a foot shear works great. Good ones can be picked up at most machine-tool auctions for a song.

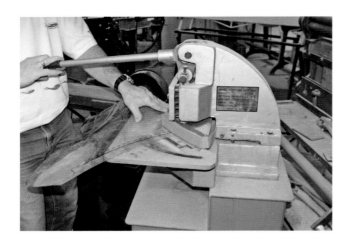

Left: *This hand-operated corner notcher will take a nice square notch of about 6x6 inches or less out of a sheet of metal.* Right: *We also have a large band saw for roughing out shapes in steel or aluminum plate. We use the band saw for rough cutting and than transfer to other machines, such as a mill or grinder, for the finishing cuts.*

oxy-acetylene torch kit is all but a thing of the past, due to the availability and affordability of the new, smaller plasma cutters.

Cutting by water or water jets is so far out of the price range of the average shop that it does not need to be addressed in this book. The technology is simply a high-pressure jet of water that is saturated with an abrasive additive that eats away the metal or whatever material is placed below the jet.

Cutting a sheet of metal with a saw is another great way to reduce the size of a sheet or modify the shape of the metal to your liking. The choice is as simple as a circular or reciprocating saw, the circular being a Skil-type saw with either a metal-cutting blade or abrasive blade mount-ed in it. Circular saws run at a slower speed than an abrasive cut-off saw. If you choose to use a reciprocating saw, be sure to check whether or not it has an orbital-motion feature. If it does, and you can, switch to the straight back-and-forth motion when cutting metal. The orbital motion is for cutting wood.

An air chisel will also cut a panel off of a car with relative ease but will generate a lot of noise in the shop, losing you however many friends you still have left. If the panel is a large one, and if you have just a small section, you could hand out complimentary ear plugs to your fellow workers first. It may be just the way to go, since it is simple, fast, and cuts through sheet metal like it was soft cheese.

Left: *A 4-foot finger brake is also very handy for making sharp, straight bends in metal, such as this faux gas tank I made for a custom motorcycle called the* Secret Weapon. Right: *The sheet metal parts, after bending, were then tack welded together, showing the final shape I wanted for the tank. Normally, the next step at this point would be to finish welding the seams, but in this case I liked the look and left it this way since it was not the real gas tank and did not have to be leakproof.*

BENDING TOOLS

Bending sheet metal requires a metal brake, a finger brake, or a press brake. The size and age of the unit determine how much it will cost, but with enough shopping around you can pick up a reasonable deal. The finger brake allows you to remove sections so you can bend a piece of metal with flanges on it, which is very common. A press brake, or forming brake, can be a very small unit or a very large unit and will allow you to make exotic bends if you have the right dies or the capability to make the right dies. A small unit will allow you to make small bends in sheet metal, and many of these units come with removable upper dies, so they act a lot like a finger brake.

A slip roll is another tool that is very useful for putting slight to extreme curves in a piece of flat metal. You could use a slip roll when you work with the curve in a driveshaft tunnel when installing new floorboards in a car. With a bit of experience, you can even perform tapered rolls, like the transmission bell housing area of the tunnel.

Each end of a slip roll is independently adjusted for the amount of curve it will produce in the sheet, which allows you to put tapered curves in a sheet of metal. The top roll can be unlatched from the entire unit, so the sheet you're working on can be removed without unrolling it. This little feature allows you to roll a tube and then remove it by simply unlatching one end of the top roll and slipping the newly formed tube off the end. Thus the name "slip roll."

Hammers are self explanatory but vary in shape, size, and usage, and you will need many of them for the different body panel shapes you will be encountering in a standard car body. There are also specialty hammers like shrinker hammers and dead blow hammers, or slapping hammers that spread the impact of the blow over a larger area.

TUBE-ROLLING MACHINES

The tube-rolling machine works on the principle of rolling a section of tube through three steel mandrels (slightly resembling a pulley) cut to the same radius as the tube. Then by an adjustment of the center mandrel downward between the two lower mandrels, a slight radius is formed on the tube being rolled between the three mandrels. The distance the top roller is from the centerline of the bottom rollers will determine this radius.

This process is one of multiple rolls, not just one, so learn to have patience, and the tube will conform to your requirements. The process is a simple one, and with this machine you can put a large radius in a long tube with ease. Or if you so desire, you can change the radius during the bend process on the same tube by simply increasing the tension on the middle roller or by running the tube through a section with a tighter radius in that area.

Tube-rolling machines can be hand powered, electric, or hydraulic. They can be small desktop machines or large floor-mounted machines.

TUBE-BENDING MACHINES

Tube-bending machines work a little bit differently than do tube rollers. They use a shoe (the curved part) and a follow bar (the straight bar). The shoe and follow bars are cut to specific diameters and can be changed to match the diameter of the tube you are bending (1″, 1½″, 1¾″, 2″, etc.). The radius of the bend is going to be constant and is determined by the shoe, such as 6″ radius or 8″ radius or whatever radius the shoe is designed to bend.

BENDING TUBES

Bending is the simple part; knowing where and how much to bend is the hard part. This is why I would suggest using

We use a mandrel bender to give us the same radius bend at each bend; we can vary the angle of the bend but not the radius with this type of bender unless we change mandrels to a different radius.

E.P. Industries makes and sells a low-cost tube roller that comes with 1, 1½, and 2-inch diameter rollers. It is hand operated but works well for small shop projects.

There are a number of software products on the market that will let you program the bend points if you know the basic dimensions, such as the length of the complete tube and the radius of the bends.

a simple tube-bending program on a computer for all your calculations, since they will take into consideration the tube diameter, radius of bend, and wall thickness of the tube, and in some you can even account for spring-back tendencies of the tube. Most tubes bend down the center-line, so the outside of the tube stretches and the inside of the tube compresses an almost equal amount. The center-line of the tube stays the same length, and this is where the accurate measurements are taken.

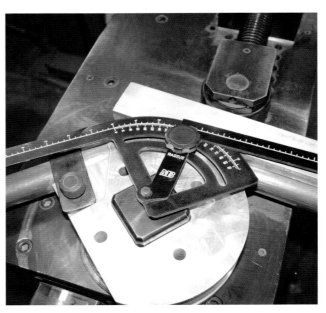

You can also buy a tubing protractor (or a set of them) to measure the amount of bend you need and then transfer the angle to the machine. Protractors are sold by radius of bend, such as 6- or 8-inch radius. They are also marked in inches of bent material so you can calculate the amount of tube used in the radius of the bend.

A digital level is also a handy tool for making sure the tube is level as you make the second and third bend, or you can purposely rotate the tube an exact amount.

Good programs to bend tube are a "must have" if you are going to take tube bending seriously, since they will give you the exact bend point from a set point on the tube.

TUBE NOTCHERS

Tube notchers are also a must if you plan on welding one tube to another and need a good, clean fit between the tubes. There are a number of good tube notchers on the market for all budgets. I used to have a mechanical notcher that used my body weight on an extended arm to notch the end of a tube with a hardened steel cutter that worked across a tray whose end was cut to match the die. Other types of notchers are small bench-mounted units that use a hole saw and an electric drill to notch the tube end. These units can be adjusted for angle and size of notch by changing the hole-saw diameter.

We also use a tube notcher that uses an abrasive belt on a rotation drum to notch the end of a tube. It is a Scotchman abrasive tube notcher and can take any size tube and put any size radius on the end of the tube at any angle. This little tool works well and is affordable for the small shop. There are also cutters that use a mill cutter to perform the same task.

TO SUM UP

Tube-bending machines are as varied as the notches are in design, and it seems every company has one. Most benders use a shoe-and-follow-bar arrangement and are generally electro-hydraulic. Electro-hydraulic benders use a 110-volt or 220-volt source to power a hydraulic pump

There are a number of ways to notch the end of a tube to fit another tube. One way to do it is cheap, but it works as long as you have the time. It uses a metal hole-saw and a fixture that can be adjusted for angles to hold the tube in place.

This is a Scotchman tube notcher. If you are doing one roll bar per month, buy the hole-saw notcher. But if you are a serious builder, this is the machine for you. It uses an abrasive belt and a matched-diameter roller to grind the exact radius you need, for perfect joints.

that fills a cylinder with pressurized oil, which in turn pushes a piston outward, driving a shoe into a rotary motion, pulling the steel tube along with it around a radius determined by the radius of the shoe.

Tube-bending protractors are for determining the angle to which you need to bend the tube. Years ago I used to use a piece of welding rod to find the angle I needed, but the rod, being small in diameter, would lose the setting at the least provocation (like being bumped by me), so I later

moved to a larger-diameter section of copper rod. I found that you can flex copper thousands of times without breaking it. So that tool (copper rod) was good for years.

Then one day I stumbled on the protractor made by Mittler Brothers, and within a day purchased one for each radius I was planning on bending. This tool can be placed in a car as you are designing the roll cage and can be rotated to the required angle and then locked at that angle for matching to the tube you are bending.

BUILDING A FIREWALL

This is a small project that shows how to build an aluminum firewall cover for a 1950 Mercury two-door coupe. Not just any '50 Merc, though. This one was used in the movie Cobra, with Sylvester Stallone. We built several of these to be crashed in the movie back in the 1980s. This one returned to us from the grave.

1 We start by making patterns. In this case, we'll use a scrap of cardboard and start cutting it to slightly larger than what is needed for the firewall. 2 Now we cut out the areas for the roll bars to slide through, and we round the corners at the top of the cowl.

3 Eventually it's time for a test fit. This pattern was cut into three pieces. Each one was fit into place and then taped together for this fitting.

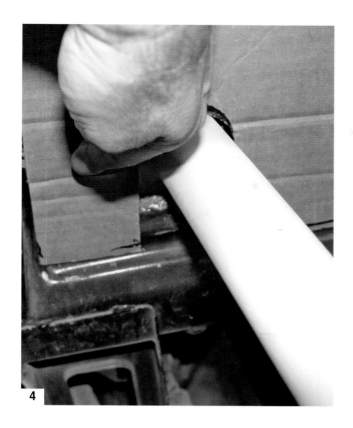

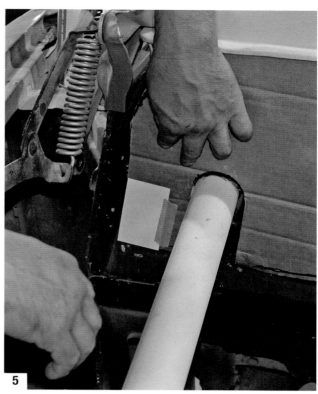

4 This fitting allows us to mark any tight spots that will need trimming. **5 & 6** A trick I learned was to keep old business cards for filler pieces. When the pattern falls short of fitting in some area, simply tape a business card or two into place. This will allow you to make precise adjustments to your pattern. **7** Simple square-cut slots can be cut and then later turned into radius cuts once the pattern is transferred to aluminum.

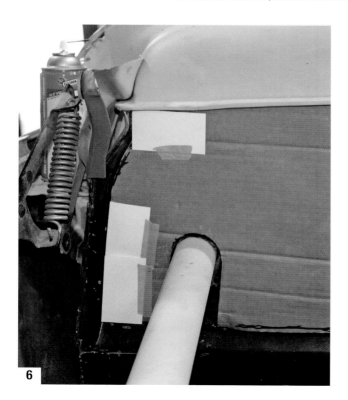

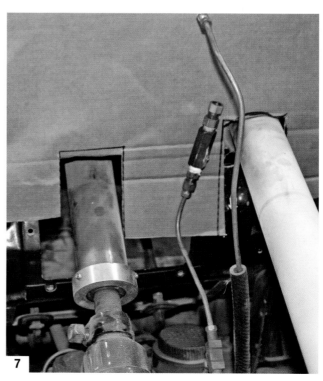

8

9

8 *The finished pattern can then be removed and taken to a workbench for transferring the design to metal.* **9** *I use a soft aluminum (3000-series) for the firewall. This will allow me to roll a bead with less chance of the aluminum cracking, as would happen if I used a hardened 6061-T6 aluminum. In this shot you can see all the business cards we used around the edges. If I didn't call you back now you know why.*

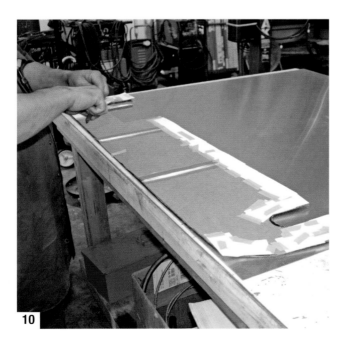

10

11

10 & 11 *Now we use a felt-tip marker to trace around the pattern. One trick I use is a revision method of different-colored pens. The original could be in red, with the first change in black, the second change in blue, and so on. This lets me know what I have changed and when.*

12 & 13 Using a set of electric three-bladed shears, we start cutting along the lines we just traced onto the metal. This is a critical cut, so take your time, or you will wind up scrapping the metal and starting all over again. 14 Using the newly cut-out firewall, we do a test fit on the car. We are looking to see if we have to bend the firewall to fit it into place. If we do, we will have a big problem after we add the beading around the edges because the beading will make the panel stiffer and harder, or impossible, to bend.

15 & 16 *As you can see, the fit was good, and we found the panel could be installed without having to flex it.*

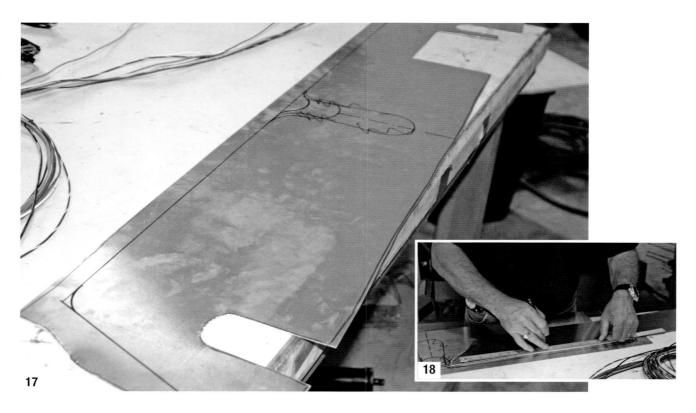

17 & 18 *The firewall is removed and taken to the workbench for laying out where the bead should be rolled. This bead will not only add a nice look to the firewall, but a lot of stiffness as well.*

19

20

19 *Using a hand-operated bead roller, I simply followed the line I had marked on the panel.* **20** *Making turns with a hand-operated bead roller is a task in itself and will normally require some help, but this is what friends are for. Just remember to crank slowly.*

21

22

21 & 22 *I had an employee turn the crank as I guided the metal through the bead roller. Later, I designed my own electric bead roller that allowed me to use a foot control instead of a crank.*

23 *A spindle sander cleaned up the edges.* **24 & 25** *Just prior to riveting the firewall into place, a last fit is in order. The firewall passes with flying colors.*

PAINT
from the Headers, Manifold Back
Part # H6202

High Temp
Coating Spray

High Temp
Coating Spray

CHAPTER 7
METAL FINISHING

The process, or shall I say art, of metal finishing is an overused term today, since few bodymen can actually perform a true metal finish without using filler. Lead was used as filler and, later, abused as filler. The abuse came about when bodymen found they could use very thick applications of lead if they needed to, and when they did, they also found it was easier to shape and work than the base metal.

Then, sometime in the 1970s, a product came out named All Metal, and body shop owners could "honestly" tell their customers that the repair was done in All Metal and involved no bondo. This filler is still around and has aluminum powder mixed in with an epoxy base. I have no idea why it is supposed to be so much better than a lot of the fillers out there today. The manufacturers claim that it creates a moisture barrier and looks like lead when sanded down.

Now, a word in defense of plastic filler. Have you ever seen an old car that was hit in the fender and the area has a chunk of body filler knocked out of it at least two inches thick? Well I have, and my first thought (besides; my God, that must be where Jimmy Hoffa is buried) was that it took a car hit to dislodge that giant chunk of filler off the fender. Now don't misread this and think I am some kind of advocate for building cars out of plastic filler. It is just that when I notice things I look for the positive in them. In this case, it is that plastic filler, if applied correctly, will outlast some lead fillers. Old cars had their seams filled with lead, and many old cars are cracking at those seams. Lead is good, but it's not perfect. The only other choice is pure steel, but a finish in pure steel is all but out of the question financially and way out of the skill range of the average bodyman.

However, it can be done, and it is well worth learning how to do, if for no other reason than for self gratification. Years ago, I used to do work for a company in Santa Monica called Automotive Classics. The spectrum of cars we worked on ranged from Bugattis to Rolls Royces to old Packards, and the owners wanted all the cars finished in metal and, once in a while, a little bit of lead. During this time I noticed that doing a fender repair using filler or lead might take about an hour. But to metal finish the fender without any filler at all would take 10 to 20 hours or more in repair time. So, you can see that the cost of a metal finish could run 10 to 20 times or more as much as

Protecting metal from the elements almost requires a degree in chemistry and a warehouse full of paints, primers, and sealers.

a fender repaired with a bit of filler. Now, there are cars and times when a metal finish is absolutely imperative, such as repairing a stainless-steel-bodied Bricklin or an old classic or even an aluminum-bodied Lotus, especially if the body is later going to be polished out to a high shine.

This Herkules paint-gun washer saves a lot of time when it comes to cleaning your paint gun. It does a better job, and it won't damage the gun, as some brushes can do if used incorrectly.

PAINT
on the Headers/Manifold Back
Part #100202

High Temp
Coating Spray

High Temp
Coating Spray

Grays are a popular color for restoration work, and there are a number of hues that can be used. Eastwood also makes PRE, which is a painting preparation chemical that not only removes wax and grease, but helps promote paint adhesion.

HOW TO PERFORM A REPAIR IN METAL ONLY

This is not as difficult as it is time consuming. You first need to use your head instead of your muscles and tap the metal instead of smashing it. You need to analyze the dent, find out how it happened, and reverse the process slowly and methodically. Start from the area furthest out from the point of impact and work you way toward the center of the contact point. In a way, you will be "ironing the dent out" by working the area around the damage.

Your primary purpose is to relieve the stress on the principal impact point. This way, it can be lightly pushed back into shape and will stay there without having the outer areas pull it back into the damaged condition. Higher-crown areas, like the front fender of an older car, are the easiest to repair. On the other hand, the flat doors on newer cars are the hardest, since the flatter panels can be overworked very easily and have the tendency to "oil can" back into the damaged condition if not repaired properly.

I sometimes spend a few hours just studying a dent, planning my strategy and order of attack, before I touch a hammer to it. I tend to use smaller hammers and lighter blows than most because the worst problem and most damage is normally from excessive stretching of the metal during the repair process.

Think of body repair this way. For every action, there is an equal and opposite reaction (Isaac Newton's third law of motion), meaning that with every hit of the hammer there is an equal and opposite reaction, where the metal moves in the opposite direction from the hammer

blow. The lighter the hammer and the softer the blows, the less the reaction, but there will be a reaction nonetheless. My concept is that, even though it takes a bit longer to use lighter hammers and lighter blows, it gives me more control over the repair, plus I will not overwork the area as quickly as I would if I used a larger hammer. To carry this point to absurdity, tapping a dent with your finger will eventually tap the dent out, but your as-of-yet unborn kids may have died of old age by the time the dent is finally tapped out.

I know guys that grab a sledgehammer and give dents a big whack, then move to smaller hammers for the finish work. They spend a lot more time finishing the area than they would have if they had just scaled the hammer down to about a tenth of what their emotions, or ego, called for in the first place. Big dent calls for big hammer, Ugggh!!! Well, there is a better way, and that is to use your brain first and figure out a strategy for removing the dent. And, again, work from the outside toward the center.

When the dent is all but out, the use of a Vixen file is in order. The Vixen file is a special file for shaping lead and steel. It has parallel cutting teeth that are like chisel blades, and they remove the high spots on the metal with ease. After a few strokes with the Vixen file, a little more hammer work will bring up the many low spots that are now easy to spot. A bit of time tapping up low spots and filing down high spots, and before you know it, the dent will be impossible to find.

The time it takes to do a metal repair the right way will be less than it would take to do it the wrong way. And, even if you do not metal-finish each dent you come across, the knowledge of how to do it will guide you toward better body work in general.

High-temperature paints are also handy for exhaust pipes and engine blocks. These are also available in spray cans.

CHAPTER 8
METAL TREATMENT

GET IT WHILE IT LASTS
FOR ONLY $19.95

Rust will reduce the value of your classic car to that of scrap metal if not caught in time and treated properly. Rust treatments have become what diet pills have become, a fast way to make a buck for a bunch of con men. The good rust treatments are buried deep in a pile of the phony rust reversals, which are the ones that claim they turn "rust" into "brand new steel" (lead into gold?) with one simple application . . . just send in your $19.95 . . . but wait there's more . . . and so the ad goes.

The process of making new steel actually takes a foundry and a lot of engineering, not an opportunist with a chemistry set who places an ad in *Popular Mechanics* offering the rust removal product of the century.

RUST, WHAT THE HECK IS IT ANYWAY?

Rust is ferrous (iron) oxide, or as his friends like to call him, Fe_2O_3. He is a compound, and his main job is to link up with the carbon dioxide (in the air) as soon as he can, then the two sit around (having a beer, I guess) and wait for a pretty water molecule to drop in so they can all get together and form a weak but effective carbonic acid. The water is used by Rusty and CO_2, the air head, as an electrolyte to dissolve the metal in your car.

As this mild acid forms on your fender and the steel is being dissolved, some of the water (possibly out of anger at having been picked up so easily by the two dudes) will begin to break down into its component pieces—hydrogen and oxygen. The free oxygen and newly dissolved steel (or, more accurately, the iron in the steel) will then bond into iron oxide, and in this never-ending bonding process it will start freeing electrons. The newly freed electrons (perhaps singing, "set my electrons free"), having just been liberated from the anode portion of the iron, will then fly to the cathode, which may be an adjoining part of the same car, or if they're feeling really, really free may even jump to the car next to yours in your garage, if it is less electrically reactive than iron. This may explain why a rusty tool will contaminate and start the tool in the next drawer to rust. It's those darn freed electrons again!

1 *I recently procured a project car, a 1940 Ford sedan, that could have been a poster car for "what rust eats." After a brief inspection, I ran out and purchased a few items to help me tear the car down. Most of these items were rust removers; some were oils to help loosen rusted bolts.*

2 *I have no idea where the grill came from, since the car was a 1940 and the grill was not. As you can see, the car did not look that bad at first glance—or in the photos on eBay, where the car was advertised as "rust free." Apparently, what the previous owner meant was that the rust itself was free!*

Unfortunately salt on roads and chemicals found in acid rain only encourage this process, and they make water into a better electrolyte or conductor, which speeds up the process of rusting on iron and corrosion in other metals, such as aluminum and magnesium.

Now, something else to consider is that hydrated iron oxide, or "wet rust," will continue to corrode the metal below the surface if it is left alone and the surface is kept moist, so the idea that rust is a protective layer is all wet in itself as well. The exceptions to this rule are aluminum and stainless steel. The aluminum oxide forms a protective layer for the aluminum, and the stainless steel gets what is known as a passivation layer called chromium (III) oxide. This is also true for copper, zinc, and magnesium, so you might want to think twice before you polish those wheels—that is a protective coat you are removing.

Well that was the bad news. Now the good news. At the bottom of the pile of miracle chemicals that turn a rusty trash can into an AC Cobra are some honest-to-gosh, real rust controllers.

Short of mounting a bunch of zincs on your car like they do on boats (a zinc is a sacrificial anode made from something with more negative electrode potential than the metal to which it is attached), you can do some of the following, which should help stave off the effects of oxidation.

CLEANLINESS IS NEXT TO RUST FREENESS

You can keep your car clean, cover it in a good coat of paint, use an undercoating that covers all the lower frame and under-body areas, and make sure that water or mud cannot collect on any part of the car. Check the drains on the door bottoms to see if they are open and will still allow water to run out of the doors. This is a common problem, because mud (containing water) collects inside doors at the bottom and starts the rusting process from the inside out. By the time you see those little bubbles in the paint, it is already too late, since the metal under the tiny bubbles is all but gone. Waxing a car also helps because it allows the water to run off the surface better. Older cars have a tendency to collect water under the stainless window trim, especially at the rear window. The front windshield collects water, which will start the rusting process if it doesn't get blown out. So, get in the habit of blowing out the water that is trapped under the chrome, and you can ward off the inevitable.

TAKE THIS RUST REMOVER AND CALL ME IN THE MORNING

If you already have rust and have to get rid of it, there are a few products on the market that will work relatively well for slowing it down, and some manufacturers even claim that they stop the effects of rust altogether (sure, you bet). Stopping rust is a relative term, though, since nothing in the way of disorder is totally stopped, according to Newton's second law of thermodynamics, which states, "everything falls into disorder rather than order" (rust falls in the category of disorder). For example, leaving your car in an underground parking structure for 150 years will not make it better, only 150 years older and rusty.

3 As I moved around the car, the rust became so apparent that it became a challenge to find a good panel. Rust tends to accumulate around seams and places where body panels meet or where mud can accumulate.

POR-15 is a product that brings the powerful technology of polymeric isocyanate derivatives to the consumer rust prevention market for the first time. According to their product information, it uses a technology vastly superior to competing products currently on the market. However, POR-15 is easier and less expensive to apply than epoxy-type coatings, since it doesn't require mixing. Also, POR-15 dries to a high-gloss or semi-gloss paint-like finish, except that it cannot be scratched or peeled off, since it actually becomes part of the surface. I have tried POR-15, and it is very good at doing what it says it does, and it's relatively easy to apply, too. I have read about a test using POR-15-treated metal and a salt spray, in which the metal resisted rust during extensive spays of saltwater. An application of POR-15 will definitely slow down the effects of rust for quite some time, but I doubt that anything will stop it forever.

Eastwood also sells a rust converter that "turns rust into a coating" that will stop, or, as I prefer to say, slow down, future rust from eating your pride and joy. We have tried and tested this one as well, and it appears to do the job for a nominal cost.

WELL THAT WAS A BLAST!

Sandblasting, grinding, and wire wheels are other ways of removing most of the rust, or at least the surface rust. They all seem to work for the removal of 90% of the rust, but they are also removing a few thousandths of an inch of the surface of the metal and distorting the remaining surface from the heat generated by the friction of the rust-removal process.

The biggest problem is that these processes do not get into the pores and pockets or between the body parts at the seams to stop the contiguous areas from spreading the rust back into the freshly ground surfaces. These types of areas are best treated by chemical dipping or by spraying some kind of rust-preventive or rust-removal product into the cracks and crevices.

The phosphoric acid (main ingredient) in Naval Jelly will also remove the rust in hard-to-reach areas as well as on the easy-to-reach areas. Just remember that when you are using Naval Jelly, do not let it dry on the metal. Keep spraying it with a bit of water to keep it working and use steel wool to rub it in. Also, don't forget the rubber gloves, unless you like burning your hands with an acidic solution. I know I hate gloves, but I hate having my hands burned more. After the rust removal is applied, don't forget the "neutralizing" of the area, or the chemical reaction will destroy you newly applied paint job.

SPRAY TO NEUTRALIZE

Some simply use soapy water to neutralize rust-removal products, but the addition of baking soda to the soapy water will be the best way to stop any chemical reaction in its tracks.

Finally, a good coat of primer on the repaired and treated metal, followed by a coat of paint as well as a good thick undercoating, will put rust to rest and allow you to relax and drive your car without having to worry about it falling apart on the road. Then all you have to do is a bit of preventive maintenance to keep it that way.

METAL TREATMENT

4 *This photo shows a rig we built that can lift the body off the frame for derusting. It has casters on each leg and two winches for lifting.*

4

5

7

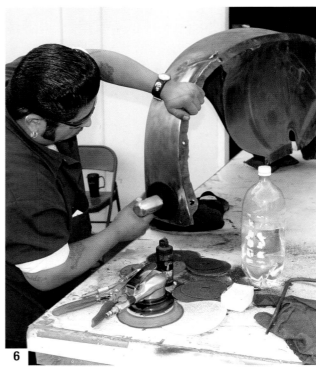

6

5 & 6 *After the car was stripped down, the parts were treated according to the surface. Some were hit with a grinder or sander, some were treated with chemicals, and still others were sandblasted.* **7** *Much of the firewall was hit with rotary wire wheels and chemicals in order to get into the hard-to-reach areas.*

8 *The floorboards were so bad that they were totally replaced, as you saw in chapter two. The main reason floorboards are often the worst is because moisture collects in the carpet padding, so the metal is seldom dry.*

9 *This is the area between the roof and the side rear body panel. It's one of the hardest places to get to, but because I was going to customize the car, it would not be a problem. All I had to do was cut out the area and replace it with a single sheet of metal. But if this were a restoration, you would need to clean up the whole area using chemicals or by sandblasting.*

10 *This area is very common for rust. Dust and dirt get trapped in the door panel, and rainwater runs down the widow, mixing with the dirt and making mud that just sits there for days and starts the oxidation process.*

11

12

11 *During the teardown phase of the project, I had to use a grinder to remove many of the bolts that held the car together. Rust had frozen them in place, and even rust-removing chemicals would not loosen them.* 12 *Using ScotchBrite pads mounted on an abrasive wheel will remove rust from most small parts without destroying them.*

13

14

15

13, 14 & 15 *The trunk is another place that gets a lot of rust, especially in the corners. We used a product called Metal-Ready by POR-15, which works very well, converting rust to a protective coating. After applying Metal-Ready, you can either apply POR-15 rust-preventative paint or another primer.*

16

17

18

16 *After rust is either removed or converted, you will need to spray a primer to seal the area. We used to do this with a spray gun but have recently started using the rattle-can system, both out of simplicity and because the product has improved so much over the years.* **17, 18, 19 & 20** *As soon as the rusted-out area is welded up or chemically treated, a body filler is applied to the area to build it back up to its previous shape. Next the filler is sanded with progressively finer paper, then it's primed. The other corners are treated in the same way.*

19

20

21

21 & 22 *Eastwood's Heavy-Duty Anti Rust is an undercoating that will be sprayed on the metal skin of the trunk but not on the bracing. Therefore, the bracing is taped off before the undercoating is applied. Anti Rust requires a special spray gun for its application. This is sold by Eastwood as well.* **23** *Before applying the undercoating, the trunk is blown off to help get rid of any dust that is under the trunk bracing.*

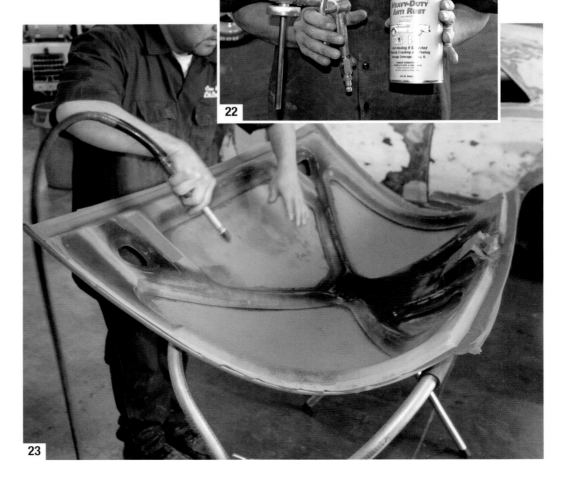

22

23

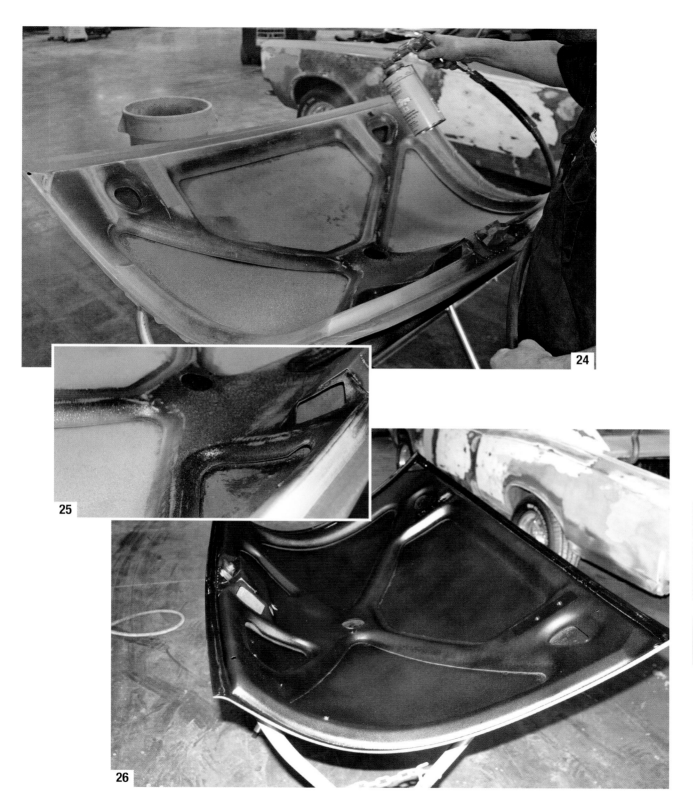

24 *Using the special spray gun, the Anti Rust coating is applied, and a thick, textured surface is left behind.* **25 & 26** *Since Anti Rust is a clear spray, a colored finish is needed. After the Anti Rust dried, we chose to use black paint. The trunk lid is set aside to dry. Later it will be fitted, and the outside surface will be painted to match the car.*

CHAPTER 9
ROLL BAR FABRICATION:
From the scribble on a scrap of paper to a complete cage and how we get there

Bending tubes can appear to be some type of black magic to the average onlooker. The way some people can measure a tube and have the bend wind up in the right place is uncanny. Even though the simple act of bending itself looks like it should be easy, all it takes is one slight mistake and the complete tube is reduced to scrap. Tube cannot be re-bent easily once a mistake is made, so you have to measure twice and bend once.

Fabricating a roll bar is quite a task if you're not armed with the right equipment and training. To that end, this chapter is designed to teach you the basics of tube bending on a real project. We'll start with the basic materials and equipment and progress to examples of fabricating a roll bar for a GTO. This particular GTO just happens to be one that was used in the film *xXx*. The owner wanted a removable roll bar for when the car is used on the streets, so it had to fit perfectly to be easily removable. The reason we have picked a removable roll bar is that it contains most of the problems you'll run into while bending metal tubes. Problems like fitting the bar within a certain area, matching angles of pre-bent tubes, angle bending repeatability, and knowing how much material will be needed, And, of course, knowing exactly where to make the bends.

SELECTING THE RIGHT TYPE OF TUBE

Before you start, it's important to choose the right type(s) of tube for the project you're building. The best type for a roll bar is seamless DOM. DOM stands for Drawn over Mandrel, a technique that keeps the inside diameter of the tube as accurate as the outside. This is by far the strongest type, since it is a solid piece of metal. When absolute strength is not required, you can use welded, or seamed, tube. A ridge left on the inside of the tube where the sheet steel is welded together can identify welded tube. Welded tube is also less expensive than seamless tube. When bending welded tube, try to keep the seam toward the inside of the curve as it bends. If this is not possible, then be sure to check the tube after the bending process to be sure it hasn't split or kinked. Another important point to remember is that tube used for automotive work is structural or

If you are going to be bending a lot of tubing, I strongly suggest getting a tube-bending program—that is unless you like to do a lot of math. These programs with tell you where to make each bend and how much tube will be required.

mechanical tube and is measured by its outside diameter. Electrical conduit and plumbing pipe is measured by its inside diameter. For example, you shouldn't use a bending shoe for 1½″ conduit or pipe on 1½″ tube because it won't fit, and if you use a smaller tube to get it to fit, it will crimp or collapse.

TOOLS

Be sure you have the right tools before starting any job. The right equipment usually means the difference between a so-so final product and a great final product. The tools we'll need for the roll bar project are gloves (lightweight gloves give better feel), goggles or facemask or safety glasses, ear protection, a hydraulic tube bender, a 180-degree protractor, a metal saw or metal cut-off saw, a

There are a number of ways to notch the end of the tube so it can be welded to the next tube. This small portable drill attachment works very well for notching a few tubes at a time. It is a bit slow, but for a small shop, it is affordable.

welder (either TIG or MIG wire-feed are by far the best), a tube notcher or metal-cutting hole saw, markers for metal, a tape measure, a laser level, a software program (I use Mittler Brothers Bend Calculator), a note pad or board for notes, and a laptop computer, or at least a print-out of the design.

The first three are the most important. ALWAYS use protective gear. Remember, you can always buy a new set of goggles, but you can't just run out and buy a new set of eyeballs.

MEASURING

It's been said that when you're getting ready to bend a tube, you should always measure twice and bend once. Try to be as accurate as possible when marking your bend

points, since they will directly affect the end result. When marking a tube, it is important that you don't do it by scratching or scoring the surface, since that could weaken the metal. Instead, use a felt-tip pen or a magic marker.

Also, remember there is a plus and minus tolerance in the software program, so you should cut on the long side of that tolerance to avoid scrapping a tube.

Level your bender, or every bend will be offplane. We built a leveling plate to ensure the accuracy of our bends. You can work without a leveling plate, but you really should do everything you can possibly do to ensure a more accurate bend.

Another note on measuring—which may seem trite but is worth bringing up, since almost everyone has done it at one time or another—when you measure the tube for marking be sure to measure it with the numbers on the tape measure oriented right-side-up to you. When in a hurry, most of us have measured with the numbers upside down out of convenience. It's only later we find we've made an error by looking at the wrong side of the inch mark, then marking the bend at 30½ inches, when we meant to mark 31½ inches. This can easily happen just because the tape measure is upside-down. Take the time to walk around to the other side of the tape.

THE BEND CALCULATOR

There are extremely handy pieces of software on the market to help you determine more accurately where you need to place your bend start points. Simply enter in the centerline dimensions and the angles you wish to bend, and the program calculates the length to each bend point automatically in both decimals and fractions. You can set the number of bends, angles, and radii; you can even calculate the approximate weight of the finished piece. In the main screen there are six different types of bends to choose from, including: simple hoop, single bevel hoop by overalls, double bevel hoop by overalls, door bars by length, door bars by angles and a custom mode that allows you to enter a sequence of as many as five bends to any custom configuration you desire.

There is also a notepad built into the program for storing notes about your bending project. Remember to copy your notes from the job to the computer so they will be stored for future use.

You can also easily calibrate the software if you find that your shoe is bending slightly more or less than what you expected from the program. It's always a good rule of thumb to add an inch or two to the overall tube length before bending because it is much easier to trim off excess than it is to add missing tube.

This page and opposite *A much better unit is the Scotchman belt notcher and grinder, with changeable mandrels for each tube size. This unit can make a perfect notch in seconds.*

BENDING MACHINES

Where many other tube benders on the market today only go to 90 degrees, the Mittler Brothers can do up to a full 180-degree bend. The Mittler Brothers. 180-degree tube bender is a self-contained, hydraulically-powered machine that makes bending tube easy and fast. It converts standard shop air into hydraulic power

It is composed of the following main parts: shoe, follow bar, saddle with pin, drive square pin, foot pedal, and digital read out. Note that the shoe, follow bar, and saddle set are matched to the outside diameter of tube to be bent. You should have in your shop sizes for the most common tube you'll be bending. Another feature that makes the Mittler Brothers 180-degree tube bender unique is the powered rotating shoe.

Bending tube with the Mittler Brothers tube bender is remarkably easy. Place your tube in the shoe, making sure to line up your marks with the end of the shoe, and pin on the saddle to hold the tube in place. Then set the follow bar on the outside of the shoe and tighten the drive screw until the shoe and follow bar are clamped firmly together. Step on the pedal and bend . . . Remember, this is a one-way process, so make sure you don't bend further than you want to go. Hit short bursts of the pedal as you get close to the needed angle to avoid overbending. Also, remember the tube will have a slight amount of spring back, and this will have an effect on your final bend.

The digital readout (DRO) should be set at zero when the bender is fully retracted. This will give consistent stroke reading for maximum repeatability. When the DRO is used in the millimeter (mm) mode, the stroke very closely approximates degrees of bend. For example, 90.0mm roughly equals 90 degrees of bend. The mm mode also displays only one position after the decimal point, which simplifies record keeping. However, check the bent tube with a protractor just to be sure of the angle.

The DRO is useful when repeating several identical bends. With its large, easy-to-read display you can see the distance of the bend quickly and easily. Don't forget to make notes of the numbers at the end of the bend. The two-inch shoe and follow bar will slightly restrict your view of the DRO, so take your time and look at the angle. All the rest of the shoe sizes will allow you a clear view of the DRO.

Always wear protective gear when working with the tube bender. It is not common, but saddles have been known to break or shatter, so goggles are a must. Earplugs help reduce the chance of long-term hearing loss, and work gloves help protect your hands.

BASIC BENDING TECHNIQUES

Take a four-foot length of the tube that you will be using and mark it at the one-foot point. The length should be as accurate as possible, since this will test the calibration bend to extrapolate data that can be used to make the rest of your bends as accurate as possible. Put the tube in the bender and perform a sample bend of 90 degrees. Don't remove the tube, but check if it's square with a carpenter's square or your Mittler Brothers protractor. To do this you have to let the pressure off so the tube will spring back slightly, giving you a more accurate angle measurement. Then adjust by bending a little at a time as needed until you wind up with a true 90 degrees. Now mark the results on the tube and keep the bent tube as a sample to eliminate having to perform the test each time you bend that particular type and size of tube—and be sure to log the information in the bend calculator or notebook.

You may even want to create a chart from your results so you can quickly determine how tube thickness, shoe size, and the degree of bends will affect the overall length of your tube. It's not necessary to measure each degree, simply record results from a 45-degree bend, a 90-degree bend, and a 180-degree bend. It can be a little time consuming at first, but in the long run it could save you hours of production.

When putting multiple bends in a piece of tube, it's important to make sure that both the machine and your tube are as level as possible. When setting up the tube bender for the first time, use a level or a Mittler Brothers Smart Tool Angle Sensor to make sure your machine is level in both the X and Y planes. If this is not done and your bender is out of level, we can guarantee that every bend after that using a level on the tube will be as well. If your floor is not level, you should consider using adjustable feet on the machine to get the bender level first. This is a simple device that can be adjusted in minutes for a perfectly level machine. We have added extra nuts to our feet to allow us to lock the bolts in place once adjusted. These feet can be re-adjusted if the machine is moved to another location.

It takes some practice to get bends to come out perfectly, so you might consider practicing on seamed tube instead of seamless, since it's fairly inexpensive at about half the cost.

ADVANCED BENDING

Now for a more complex example, we will illustrate some of the features of the Mittler Brothers 180-degree tube bender during the construction of a roll cage. Start by formulating a plan or sketch of the intended cage. This can be as simple as a sketch or as complex as a CAD drawing with dimensions.

We start the job by taking the required measurements for the roll bar, in this case the major width of the car interior and the height we want the bar to be. The idea is to keep the bar as close to the inside of the car as possible so it doesn't get in the way if the back seat is to be used. This will require a few extra bends about midway up the bar. We are using 2″ tube, and the bend protractor for the 2″ tube is put in the location of the soon-to-be roll bar. This simple tool is worth its weight in gold because it will give you the angle and show you the inside as well as the outside of the tube. This way we can check for seat clearance once the bar is in place, eliminating surprises. Once all your measurements are taken, make a sketch or just take it to the software program and punch in the numbers. Just for fun, find out the weight of the tube.

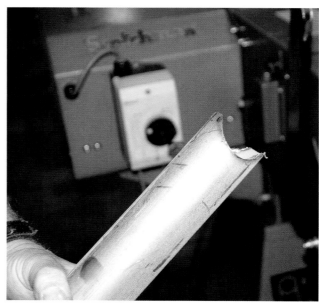

SQUARE TUBE

Mittler Brothers also offers a shoe and follow bar for square tube, and it's as simple to use as the round tube bender shoe and follow bar. The only difference is that the square tube will expand a bit, requiring you to tap the tube with a rubber hammer to pop it out of the follow bar or shoe. This will not damage the tube and is a slight inconvenience compared to the results you will get.

OVERVIEW

As you can see, there is no real black magic behind tube bending—it's merely paying attention to details. With a little time and practice, you'll be making perfect bends almost every time. With the Mittler Brothers 180-degree tube bender and bend calculator software, it's even easier to achieve that goal. A good fabrication shop really isn't complete without it.

You can perform a 180-degree with almost any other bender if you relocate the tube after your first 90-degree bend, but try to do this without your saddle marking the tube or kinking it. Then you'll see a 180-degree bender is a machine that stands alone as a basic tool for the serious fabricator.

Above: *One of the Scotchman's big features is that you can tilt the clamping table to just about any angle, allowing you to make joints between any size tube at whatever angle is required.* **Below:** *Tube bending is an art you will need to practice, but once mastered it will allow you to make intricate bends resulting in a very professional-looking product, such as a frame for a larger motorcycle.*

ROLL BAR FABRICATION

104

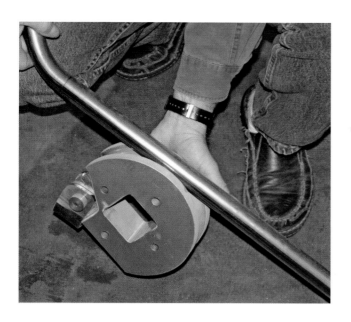

Left: *Most tube benders require a shoe, or mandrel (shown in photo), and a follow bar (straight bar with a radius cut into it). These are made to fit specific diameters of tube, so a few sets will be needed if you plan to bend different tube sizes.* **Right:** *As I said, you will need different sets of mandrels to match the different sizes of tube you will be bending. If you have a mill, you might be able to make you own. This is what we do. The most popular sizes we have are 1 inch, 1¼ inches, 1½ inches, 1¾ inches, and 2 inches, both round and square.*

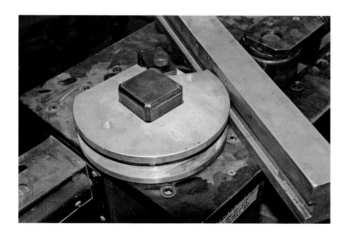

Above left and right: *This is a tube-bending machine without the tube in it to show the relationship of the shoe to the follow bar. A hydraulic cylinder under the table rotates the square pin, which rotates the shoe, pulling the tube around its radius. The mandrel and follow bar shown in the photo are milled to fit square tube.* **Left:** *This unit uses an air-powered hydraulic cylinder. The roll bar in the GTO in the background was made in about one hour on this unit.*

Top left: *This is a bar and tubing bender that will bend ¼-inch bar and tubing to 90 degrees. It is an electro-hydraulic unit requiring 110 volts to power the hydraulic pump.* **Top right and above left and right:** *These units are tube-rolling machines that can put almost any radius into a long tube by using a system of three wheels. The center wheel is adjustable up and down, and the two outside wheels are the drivers. As the center wheel is adjusted downward, it determines the radius at which the machine will bend the tube. With one of these you can roll a complete circle or even a spring (large diameter). If needed, you can vary the tension on the center wheel during the bend and vary the shape of the curve from one end to the other.*

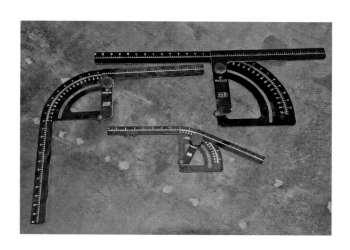
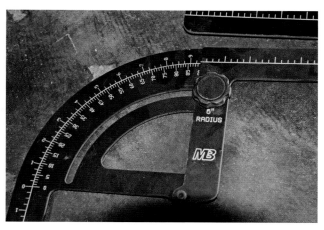

Tube protractors are handy for determining the angle you need to bend. They come in different radiuses to match the tube-bending dies. Simply hold them in place and adjust to the correct radius.

As you can see, the roll cage is the last thing you want to add to your car, unless you just like climbing over and through tubing to work on your wiring.

CHAPTER 10
SHEET METAL PROJECTS

The best way to start a metal project is with a plan, a drawing, and a list of materials, as well as a few tools. The drawing allows you to visualize the outcome of your labor. Then think about the project for a while, but not just in a positive way. Think negatively about it. That is to say, think about why it may not work, or why it may not work well, then work these problems out. If you are a good designer, the questions you ask yourself will seem endless. To illustrate the design process, we'll build a motorcycle gas tank in this chapter.

QUESTIONS

Let's start with some basic questions.

Will the fuel volume be enough for the engine? Will the fuel flow freely to the petcock valve when you ride or will there be some trapped in a pocket somewhere? Is the tanked vented? Does it have vibration damping so it will not crack? Will it be leak proof, and how do you test that before you add gas to the tank? Does it flow down to the carburetor using gravity or do you need a fuel pump to get it to the carburetor? Will it look good with the bike? Or does it stand out in a bad way, or look like and "add-on"? Well you get the idea. There is a lot more to building a gas tank than just building a gas tank.

Then there are aesthetic questions. These are the foremost reasons for making a new gas tank in the first place. After all, unless you are building a bike from scratch, why replace the tank at all? Just run the stock tank. Unless the stock tank falls short of your aesthetic expectations.

Once you decide to build it yourself, your design is what you need to concentrate on. Will the design flow with the overall look of the bike or will the tank stick out like a sore thumb? Design, of course, is open to debate, but there are norms a civilized society can agree on.

There are practical considerations. The tank of a motorcycle is for holding and distributing fuel and should be the most economical shape for holding the maximum amount of fuel with the least amount of surface area. Round or convex panel shapes are much stronger than flat panels, and the longer and flatter the overall design the less fuel it will hold, so your imagination will be a bit limited by simple solid geometry. The tank well will also limit the amount of fuel you can hold in the tank, so make it as tight to the top frame rail as you can. Use that precious space for gas, not air.

Do you have everything covered in your design? How are you going to mount the tank? Will the mounts look good? Will they be easy to reach so you can remove the tank when you need to (notice I said "when," not "if")? How about the gas cap? Is the gas cap at the highest point of the tank when the bike is on the kick stand or will you have to tilt the bike up each time you fill the tank with gas? Is the cap a new design or a retro kit that you welded in? Does it have a vent that will take the gas to the ground if you overfill the tank?

How about contiguous parts? Did you take into account the front end's turning range and the possibility of hitting the gas tank with the top triple clamp or the handlebars? Don't laugh, this is actually a good reason to visit the chapter on mockups. I knew a guy who limited his front fork's turning radius because he mounted his gas tank too far forward. He was constantly making ten-point turns in his garage because he could not turn his front wheel far enough. But he liked the look of his tank, and it was his bike.

Does the tank sit too far back into the seating area? This is also a problem easily solved by making a mockup. Can you reach the petcock while riding or will you have to pull over and get off to get at it?

Mounting the tank to the frame is not as simple as just welding on a set of tabs or even welding the tank to the frame, which, incredibly, I have seen. You need to consider the vibration of the engine and the movement of the tank on the frame as you hit bumps in the road. This constant vibration and bumping can and, in time, will cause small cracks to develop in the tank-mounting tabs or even in the welds of the tank itself, leading to the flammable, corrosive contents spilling onto your custom paint and worn-out jeans. As for the guy who welded his gas tank to the frame, well, he had to tear his entire bike apart to fix a small leak in the tank. He then had to repaint the entire bike.

Rubber mounting the tank will help stop vibration cracking but will add a layer of complication to the design.

TIME

The next consideration is time. You need time to complete the project. This is a main consideration when building any project. We are all rushed and want everything *now*! Instant gratification is the idea of the day. If this is what you want, buy a faster chip for your computer instead of building a gas tank, since there is no way to rush this job without messing it up. You will be committed for a length of time you may not have, and the worst thing to do is to start something that you will not finish. Be sure you have the time to work on the project until it is finished, even if this only means setting aside a small block of time each day.

There was a TV show that I liked as a kid, but instead of watching it, I spent the time building wooden ship models. I now have about ten models that are priceless to me and do not even remember the name of the show that I missed. So the time is there, you just have to set it aside and organize your priorities.

BUDGET

Every project has a budget and, depending on the magnitude of your project, you may need to set the funding aside in advance. The worst thing you could be forced to do is stop a project for lack of funding. Once you stop, for any reason, it is hard to restart a project. Look at the unfinished car or bike in your friend's garage. No matter the reason he stopped working on it he *did* stop, and may never finish. These kinds of projects are nothing but "volume grabbers," taking up valuable shop space.

Next, you need to make a list of materials you will need for the project. I try to buy the materials up front. That way, I won't have to wait for a delivery or for a store to open to finish my project. I will only have to stop to eat when I am hungry and rest when I am tired. I want my materials to be "in house." I even check the list from time to time to be sure I have not overlooked anything. In addition to materials, my list will include tools I might need in order to do the job and supplies like gas for the welding tanks.

MATERIALS

The type and quantity of metal you choose is as important as the design. Do you want to make the tank out of steel or aluminum? Aluminum is much easier to shape but requires more skill to weld. Steel, on the other hand, is easier to weld but harder to shape. Steel is stronger and will not start to corrode by electrolysis from being bolted to a steel frame that carries the electrical system's ground. Also, most gas caps and mounting flanges are made out of steel, so that could be the determining factor. Having fabricated both types of metal, I think you should not only be adept at working with steel and aluminum, but learn to try other options, such as copper or brass. The only way you'll develop new skills is by experimenting, testing your abilities to learn new welding and forming techniques.

Finally, don't forget the expendables, such as welding rod, filler, and sloshing compound. Sloshing compound is a substance for sealing gas tanks after welding to guarantee they will not leak. It is a heavy sludge suspended in a resin or epoxy that fills small holes and remains impervious to petroleum products like gasoline.

A FEW CLOSING DESIGN THOUGHTS

I cannot stress enough how important it is to constantly fit any part you are building to its adjoining parts throughout the fabrication process. Almost everything you do will affect the fit by slightly changing the shape. Just a small weld will change the overall shape of a tank, maybe moving the mounting holes a fraction of an inch. Remember the discussion at the beginning of this book about compounding error? Four or five "little" welds, and the mounting holes on the tank might not line up with the mounting holes on the frame.

Form follows function, or so I am told, but with some of today's designs this no longer applies. Many builders start with a shape (form) and try to make it work (function). This can get them in a lot of trouble down the road when the design interferes with how the thing they built works. A great-looking sketch cannot always be built. I often say you can draw an optical illusion, but just try to make one.

I once built a few bikes for a film. They had to be futuristic and they had to be made out of BMW motorcycles. The drawings looked pretty cool, but the design was too narrow to fit the bikes, and the end result was much wider than the artist wanted. He could not understand why I could not put a narrow body on a fat bike. Compromises often have to be made in design, so be prepared to make changes throughout the course of a project. Consider the sketches as a guide, not the holy grail of the design.

MAKING ALUMINUM INTERIOR TRUNK PANELS FOR A '50 MERC

1 The first part of any project is to make a pattern out of something like cardboard or posterboard or foam core. This pattern gets transferred to the sheet of, in this case, aluminum that I will be working with. 2 Using an electric hand shear, the panel is rough-cut into shape. Remember two things about using electric shears: one is that you should only use them for rough cuts; the other thing is when cutting aluminum, use a spray oil like WD-40 on the metal, or the blades will bind and cause your tool to stall.

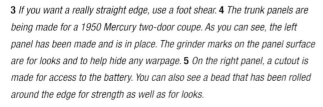

3 If you want a really straight edge, use a foot shear. 4 The trunk panels are being made for a 1950 Mercury two-door coupe. As you can see, the left panel has been made and is in place. The grinder marks on the panel surface are for looks and to help hide any warpage. 5 On the right panel, a cutout is made for access to the battery. You can also see a bead that has been rolled around the edge for strength as well as for looks.

6

7

8

6 *Sometimes I have found that laying down tape will give you a relativity straight line to follow when cutting out a center section, and it keeps the saw or shear from scratching the aluminum.* **7 & 8** *You can grind designs, names, or just a pattern into the surface and then either anodize or clearcoat the aluminum for protection against the elements.* **9** *Learning from the firewall bead-rolling tribulations detailed in chapter six, I designed and built my own bead-rolling machine with a foot-operated electric drive and a very deep throat.*

9

10 *I use ¾-inch tape as a guide for rolling the bead, since it is easy to follow and can be changed with ease.* **11** *Each panel is just set in place until the cutting is complete, allowing you the room you need for making patterns for the last panels.*

12 *With the front panel finished and in place, the trunk area is looking better. You should work from the biggest panel to the smallest panel, in that order.* **13** *If done correctly, you will not be able to see the seams. This entire job only took about four hours and made what would have been otherwise ugly trunk into a thing of beauty.*

FABRICATING SUNKEN TAILLIGHTS ON A 1950 MERCURY TWO-DOOR COUPE

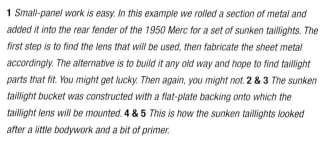

1

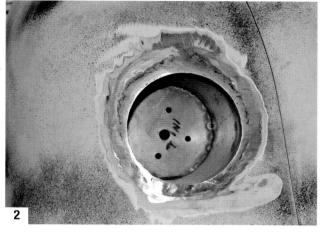

2

1 Small-panel work is easy. In this example we rolled a section of metal and added it into the rear fender of the 1950 Merc for a set of sunken taillights. The first step is to find the lens that will be used, then fabricate the sheet metal accordingly. The alternative is to build it any old way and hope to find taillight parts that fit. You might get lucky. Then again, you might not. **2 & 3** The sunken taillight bucket was constructed with a flat-plate backing onto which the taillight lens will be mounted. **4 & 5** This is how the sunken taillights looked after a little bodywork and a bit of primer.

3

4

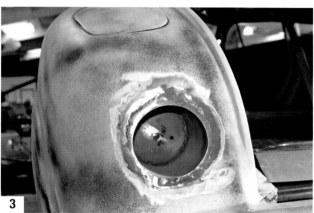

5

SHEET METAL PROJECTS

113

BASIC METAL REPAIR AND PATCH PANELS

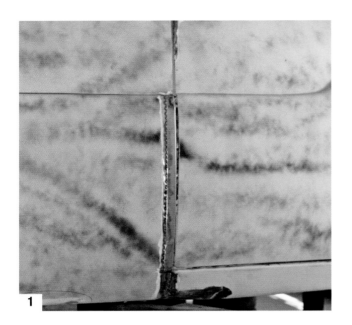

1 *This is a simple "patch panel" we fabricated to fix the front edge of the Merc's right door frame, which had rotted away. This kind of work is easy and really takes no time at all but will make the difference between a 100-point car and a 90-point car at a show.* **2** *Rockers are notorious for rust and rot as well as curb damage. Here, a patch panel was made and welded into place, and the car was back to stock.*

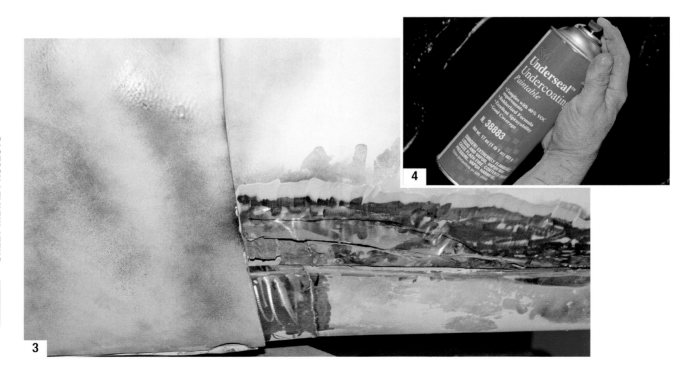

3 *The front bottom of the left door was rusted away, so we cut out the corroded metal, fabricated a sheet metal patch, and welded it in. With all rust repair, be sure to remove all the corroded metal, or the problem will reoccur.* **4** *Once you are finished with a panel, you should undercoat it, not only for looks, but for protection against moisture and future rust.*

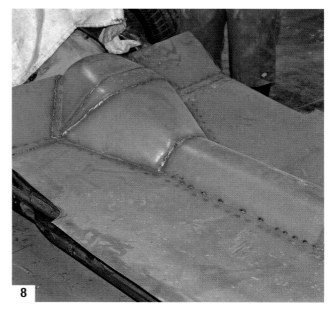

5 *A car lift is worth its weight in gold. It allows you to reach any part of the car with ease. This will make a noticeable difference in the quality of your work.* **6** *Most of the cars we do are frame-off restorations. The bodies are removed from the frame and worked on separately. In this way, we can reach everything on the car.*

7 *Do you have any idea what a grille for a 1950 Merc is worth? That's why this one, as bad as it was, had to be repaired, rewelded, and then rechromed. We just took it one bar at a time until it was complete.* **8** *Even though it may be considered a major project, I consider floorboards one of the easiest and most rewarding jobs you can take on when restoring a car.*

9 *We had to add larger tires to the front of this Merc, which involved cutting and moving a section of the fender back two inches and then fabricating a small piece of filler metal to fill the gap.*

10 & 11 *Filling in turn signal openings, as well as reshaping the grille opening of a car, is a common job around our shop and requires a bit of metalworking skill, but not as much as you would think. As long as you break the job down into small "bite-size" chunks, it is manageable by even the novice.*

FABRICATING A FOUR-LINK REAR END FOR A 1940 FORD

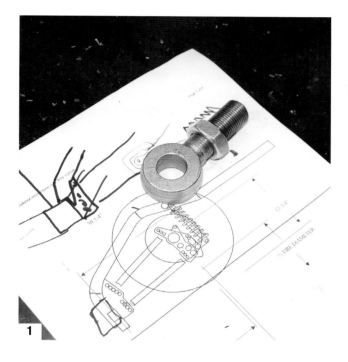

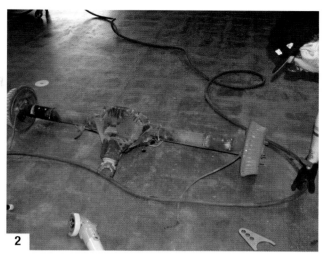

1 I like to start any new project by making an AutoCad drawing. It helps sort out the fabrication process. Then I start ordering all the parts I'll need. **2** I removed the rear end and cleaned it up in preparation for welding.

3 The axle was ground to bare metal where the gussets will be welded into place. **4** The gussets were designed and downloaded for cutting. I used ¼-inch steel plate.

5 The gussets were welded onto the axle after being aligned to the angle I wanted. 6 The two gussets were made with a split at the axle hole, then they were tack welded into place. Never weld them, or pretty much anything else, completely until the job is finished. That way you can grind them loose and adjust them if you need to.

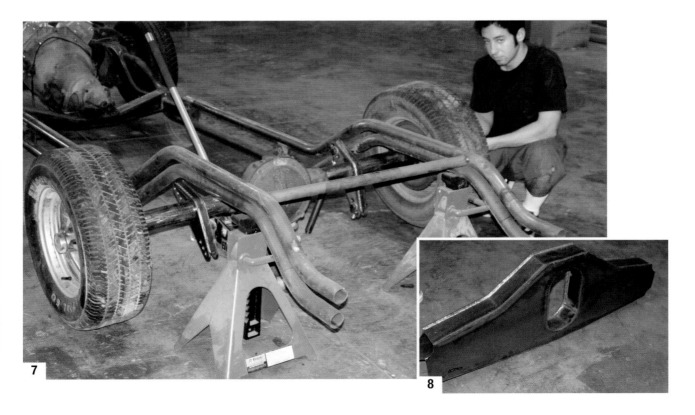

7 The rear-axle assembly was placed under the frame for a quick check of clearances and fit. 8 Now a frame support and crossbrace was fabricated. It will be a good location for the four-link's front mounts.

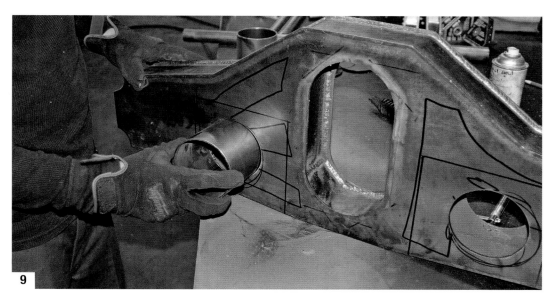

9

10

11

9 *The crossbrace also fitted with holes for the driveshaft and tailpipes. The tailpipe holes have tubes welded through them to fill in the space and add strength.* **10 & 11** *Sure it is overkill, but a ¼-inch-thick plate was added around the hole on one side to increase stiffness.*

12

13

12 & 13 *A stiffener plate at each end makes the unit real close to bulletproof. One big reason for all the stiffening is for side protection in case the car ever gets T-boned.*

14, 15 & 16 *Now the unit is added to the frame and aligned for welding. After it is tacked into place, it is time to locate the brackets to which the front ends of the links will be mounted.*

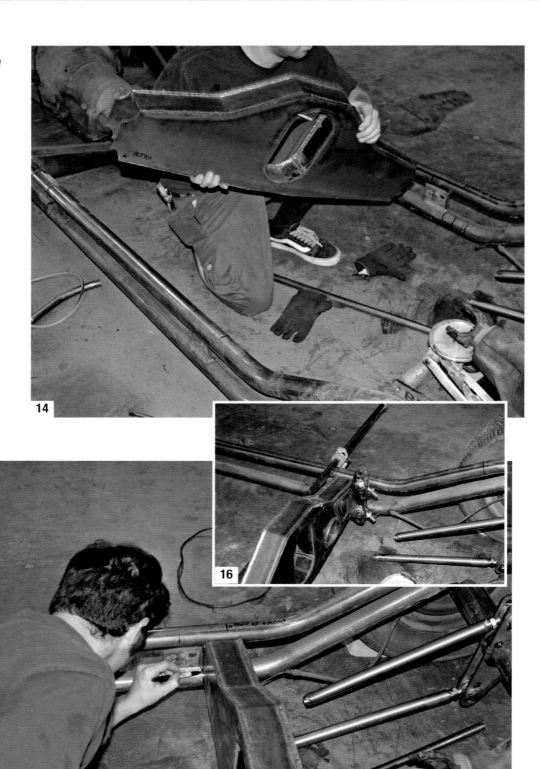

14

16

15

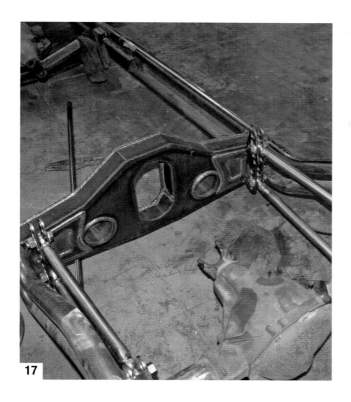

17 & 18 *The rear crossmember is now part of the frame, with adjustable front brackets for the four links.*

19 & 20 *Mounting brackets were added to the axle for the bottoms of the rear shocks. Another crossbrace was fabricated and welded into place to carry the mounting brackets for the tops of the shocks.*

21, 22 & 23 *The Panhard rod was measured and mounted to a set of custom-made brackets. One was welded to the crossbrace and the other to the rear axle.*

21

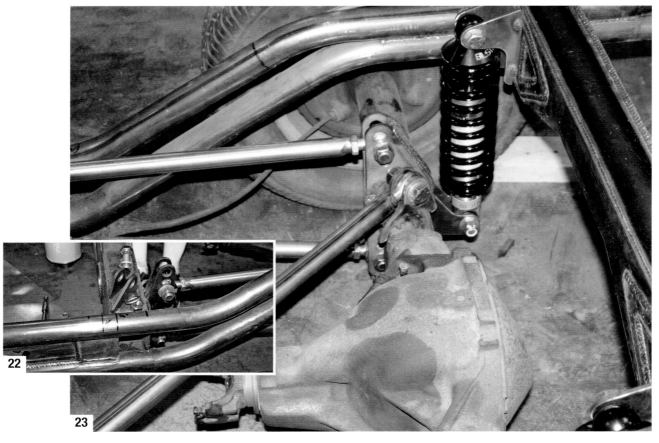

22

23

24 *The next thing to do is run the exhaust pipes through the crossmember to the new mufflers.*

24

25 *I like to set the body on the frame from time to time to check the overall look and discover any possible obstructions from the newly added hardware.*

25

FABRICATING FRENCHED HEADLIGHTS

1, 2 & 3 *To make a set of Frenched headlights, I always start by choosing the lamps. Then I use a plastic vacuum-forming machine to make the headlight buckets.* **4** *I made a set of headlight-mounting plates on a mill.*

5

6

7

8

5 *The bulbs are given a test fit in their newly fabricated buckets.* **6 & 7** *A 1940 Ford front end is much like the 1939 front end once the headlights are removed.*
8 *Both mounting plates are held in place for measuring purposes.*

9 & 10 *A crossbar can be clamped to the mounting plates to help you make fine adjustments.*

11 *An idea came to me: A section of exhaust pipe cut along its long axis would make a nice leading edge for the new headlight housings.* **12** *To cut the tube in half, I mounted an angle plate to the bed of a large band saw and used it as a guide.*

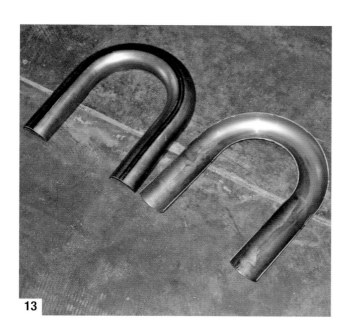

13 & 14 *The idea worked! These will be the leading edges of the new Frenched headlights. Your imagination is your greatest tool—use whatever you can for whatever you need.*

15, 16 & 17 *By cutting more pipe in a different direction and welding the pieces together, I can finish the edges with bottom returns.*

18 & 19 *The new units are welded to the bottom of the front fender, tilted to the proper angle, and a top rod is added.*

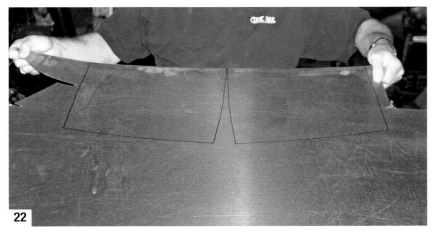

20 *Using clear thin plastic sheet to make a pattern, I bend it over the top of the fender and mark the cut lines that will be transferred to the sheet metal.* **21 & 22** *I mark and cut the pieces for the left and right sides at the same time using one pattern. Both sides should essentially be the same, just mirrored.*

23 & 24 *Using a slip roll, the metal plates are rolled and checked for fit until the metal is the exact radius I need.*

25, 26 & 27 *When welding the sections into place, first tack one side and then the other, so each weld has time to cool. This minimizes warping.*

28 Fitting the parts is essential if you want the final product to look perfect. Each weld will affect the shape of the whole unit, so constantly check the fit of the headlight components. **29 & 30** An English wheel shapes the side panels to fit the body. After they take their proper shape, they can be welded on.

31

32

33

34

31, 32, 33 & 34 *As you weld, you will need to shape the metal with a hammer and dolly. The goals are to keep the edges flat and to align the metal for welding.*

35

36

35 *When completed, the welds should be ground flat using a small air grinder.*
36, 37 & 38 *It is now time to remove the fenders so the inside can be worked and the overlap cut off.*

37

38

39 & 40 *Inner fender fillers can be made and added. The fenders can now be ground to shape and bodywork started.* 41 *The headlights are once again checked for fit.*
42 & 44 *Frenched headlights can change the whole look of a car. The amount of time required for the job is about two days, assuming you have the equipment on*

MAKING METAL SKIRTS

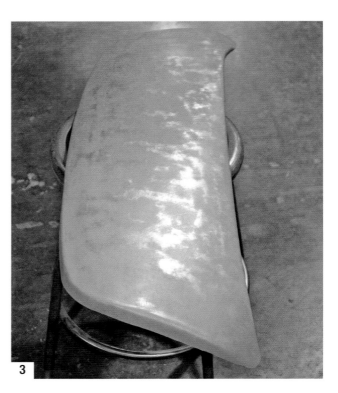

1, 2 & 3 *These skirts were formed between two pieces of plywood that were cut to the shape I wanted. The metal was hammered around the edge of the plywood to make the edge of the skirts.* **4 & 5** *The metal is worked between a hammer and a dolly to work it into the proper curve.*

6

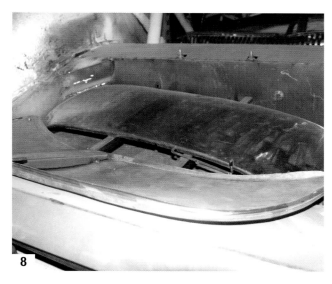

8

7

6 & 7 *The skirts were fitted and trimmed about 20 times until they fit like a factory skirt would.* **8** *Mounting brackets were added to the backside of each skirt. Then the skirts were ready for paint.*

9 & 10 *Primer was sprayed, and the skirts were block sanded to perfection. Then they were painted to match the car.* **11** *This photo shows the skirt without any filler, proving that a metal finish can be achieved if the body man knows how to shape metal.*

9

10

11

EXTENDING A MUSTANG HOOD

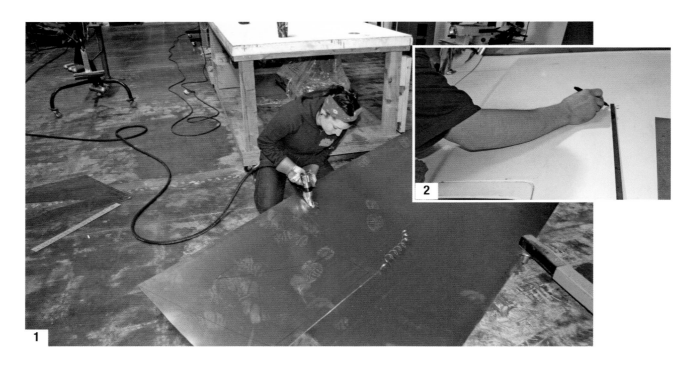

1 *It will take about one sheet of metal to make this hood extension. Use the same gauge steel as the car's existing body panels.* **2** *What I wanted was a hood extension that would change the whole look of the car with little effort. I started by finding and marking the center of the hood for a reference point.*

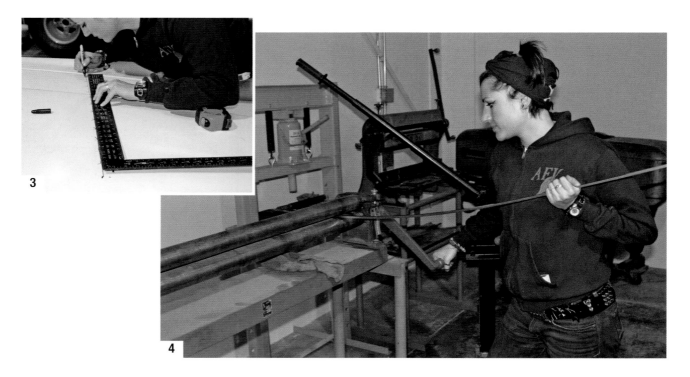

3 *From this reference point, I then have one of my employees mark every two inches from the center to one of the outside edges of the hood.* **4** *A ¼x½-inch bar is bent to the curve I want the new sheet metal to follow. This will be used as a reference during the layout.*

5 *The bar is then placed against, and in the center of, the hood to see how it looks. This is the last chance to change the curve if you do not like it.* **6** *A trick I use to find the contact point between the hood and the bar is to slip a business card under the bar and slide it until it sticks. The point where it sticks is the contact point at the high point of the hood.*

7, 8 & 9 *By moving the bar from the middle of the hood to one of the outside edges along the previously marked two-inch increments and sliding the card to find and mark the high spots, you will end up with a map showing where the metal should be cut to match the crown of the hood.*

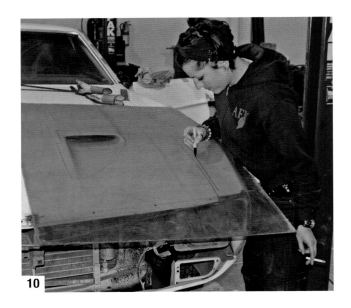

10

11

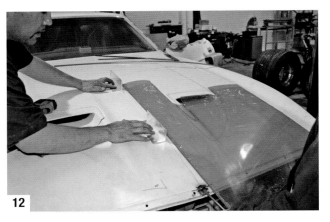

12

10 *A clear piece of plastic is laid on top of the hood. This allows you to see the marks that were made on the hood so a good pattern can be marked on the plastic to be cut out and checked.* **11 & 12** *Cutting out the pattern and fitting it shows what the shape of the metal will be. This shape is now transferred to the metal.* **13 & 14** *Both sides are cut with the same pattern and then taken over to the roll where a slight curve is added to the sheets to match the bar that we shaped earlier.*

13

14

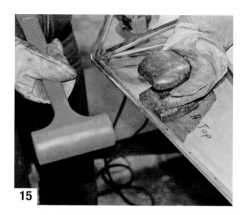

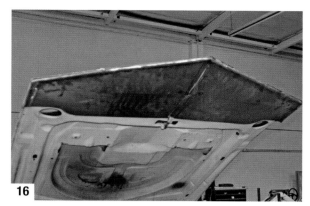

15 & 16 *A front bottom filler piece is added using the same process shown above. The front edge is bent and shaped up about ½ inch to meet the new hood extensions that will be welded to the hood.*

17 & 18 *After welding and a bit of shaping, the new look can be seen.*

19 *Don't forget to cover the windshield, or the welding process will ruin it. The sparks like to stick to glass, which will reduce a windshield to rubble.* **20 & 21** *Not exactly a 1972 Mustang anymore is it? Yeah, yeah, so I got a little carried away. I can't help myself—I am a customizer!*

SPLIT-FRONT WINDSHIELD DIVIDER

1 *I came up with this idea when I chopped the top of my '40 so low that I had no back window. I hated to throw the rear window frame away, so I decided to use part of it on the front windshield as a divider.* **2** *The center divider was the part I wanted, so I cut it out with a saw after marking the cut points.*

3 & 4 *The divider had to be reduced by about 6 inches to fit the window opening.*

5, 6 & 7 *Even small parts require a lot of trimming and fitting in order to look as if they belong there.* 8 *On the front windshield, the frame has a peak. This had to be added to the new splitter by using a hammer and a vice.*

9

10

11

9, 10, 11, 12 & 13 *A lot of cutting, trimming, grinding, more cutting, some snipping, more grinding, and a bunch more fitting, and I was almost there.*

12

13

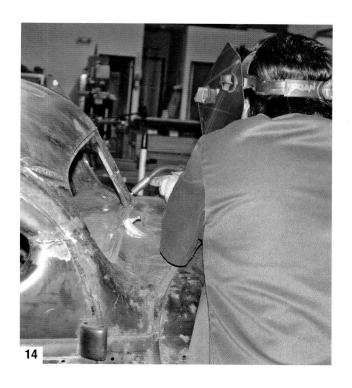

14

15

14 & 15 *It was metal melting time as the three parts became one again.* **16 & 17** *The divider came out as expected, delicate but not too subtle, bold but not abominable. But what a lot of work for such a slight change.*

16

17

CHOPPING A TOP

1 I found a new way to chop a top using a laser to mark the top. As the laser beam wraps around each post, it allows you to mark it as level as possible. You will need a laser that produces a straight horizontal line. 2 First you will need to level the car from side to side and front to rear.

3 Raise the laser to the proper height, turn it on, and use the line it produces on the car to mark the cut lines. (You can't see the laser line here because it doesn't photograph well. Usually you need to wear special filter glasses to be able to see it on the jobsite.) 4 Each post is treated as if it were the only one being cut. In other words, the cut lines do not have to be level with one another from post to post—as long as you take out the same amount of metal from each post. What you are looking for is a section that, when cut out, allows the top of the cut and the bottom of the cut to realign when the top is dropped.

5 Here the C and D pillars are marked at different heights, but the same length of metal is to be removed. This way we can remove the hinge from the C pillar and replace it later with a hidden hinge. 6 You can see that if this section of the A pillar is cut out, the upper part of the pillar will not align with the bottom part unless we move the whole top forward. To do this, we will be cutting across the top from side to side.

7 & 8 On the D pillar, the marks wrap around and basically remove the rear window. That's what you will get with a six-inch chop. You can always mount a video camera so you can see when you're backing up.

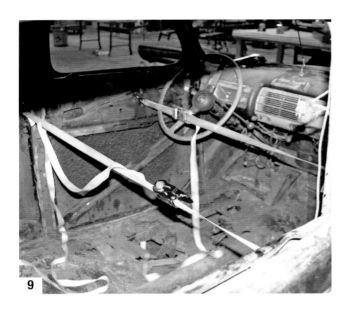

9 *Before the chop job, strap the car together so it will not distort as the top is cut off. I use tie-down straps for this.* **10** *Now each of the doors is taped together with duct tape to keep them from opening during the cutting process.*

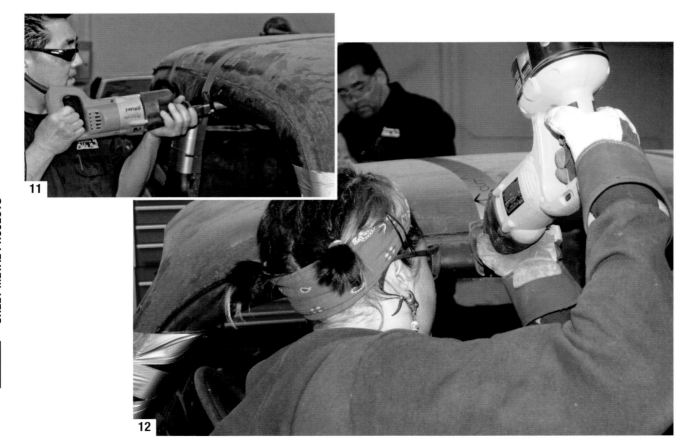

11 *And the first cut is made across the top between the "A" and "B" pillar. This will allow us to align the A pillars as discussed earlier.* **12** *We cut the top from both sides toward the middle at the same time, for no reason other than to speed up the process. The whole chop job took about two days. So I guess it helped.*

13 & 14 *We attacked the A pillars, and within minutes the top was severed. Then we made the top cut on the doors.* **15 & 16** *The remaining A pillars are cut down to the proper height so the front section of the top can be reinstalled at the new, lower location.*

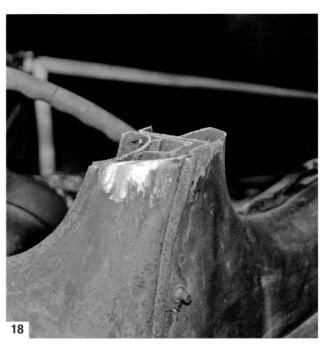

17 *Measure the front part of the top to make sure it is parallel with the door top.* 18 *The front of the top is removed, and the post is ground for a better fit.* 19 *The post is a bit narrow and will have to be widened by slitting the roof near the top of the A pillars and pulling the A pillars outward to fit.* 20 *Once the top fits, it is tack welded in place.*

21 & 22 *Now we move back one section and start the whole process over. You can barely see the slits at the outer top of the windshield that allowed us to widen the tops of the A pillars to fit the bottoms of the A pillars.* 23 & 24 *The B pillar section of the top is removed and set aside for trimming.*

25 The real test of the new top height is the "look out the windshield and see how you look" test. He looks cool! 26 Again the welding begins and the B pillar section is tack welded into place. The gaps across the top will be dealt with later. 27 & 28 The B pillar section is starting to show the final result. The gap between the A and B pillar sections is exactly what we predicted it would be: 3½ inches.

29 & 30 The rear window is cut out and saved, since it will no longer fit in this car. 31 The A pillar is welded up a bit more so the front section will stay in place better. 32 The B pillar is checked for fit and will be trimmed a bit before it's welded into place.

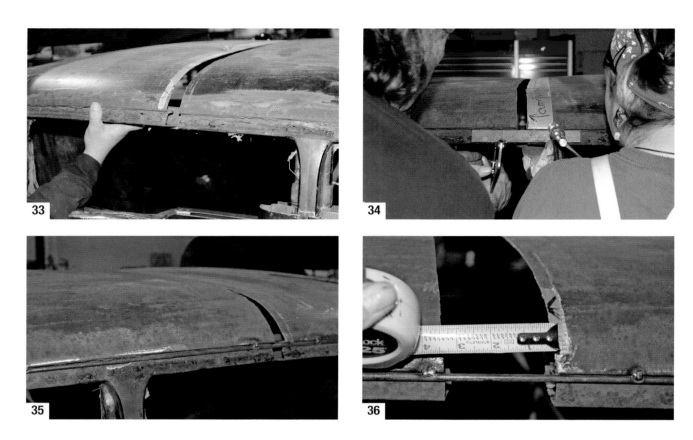

33, 34, 35 & 36 *It is time to start aligning the gap between the A and B pillar sections of the top. You can do this with a steel bar and clamps or by tack welding a round ¼-inch bar. This is simply to hold the alignment of the two roof sections while the rest of the top is chopped. The bar will be removed later.*

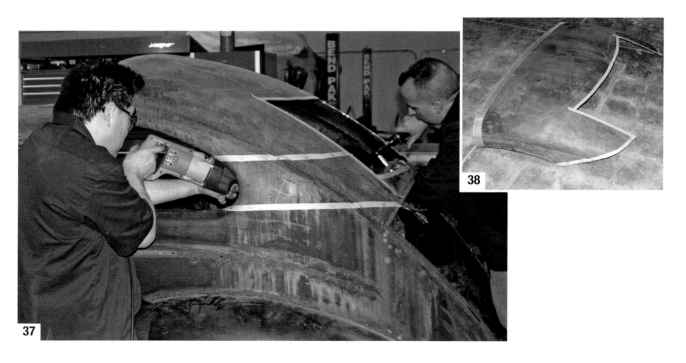

37 & 38 *Now the last part of the top is cut off and laid aside for trimming.*

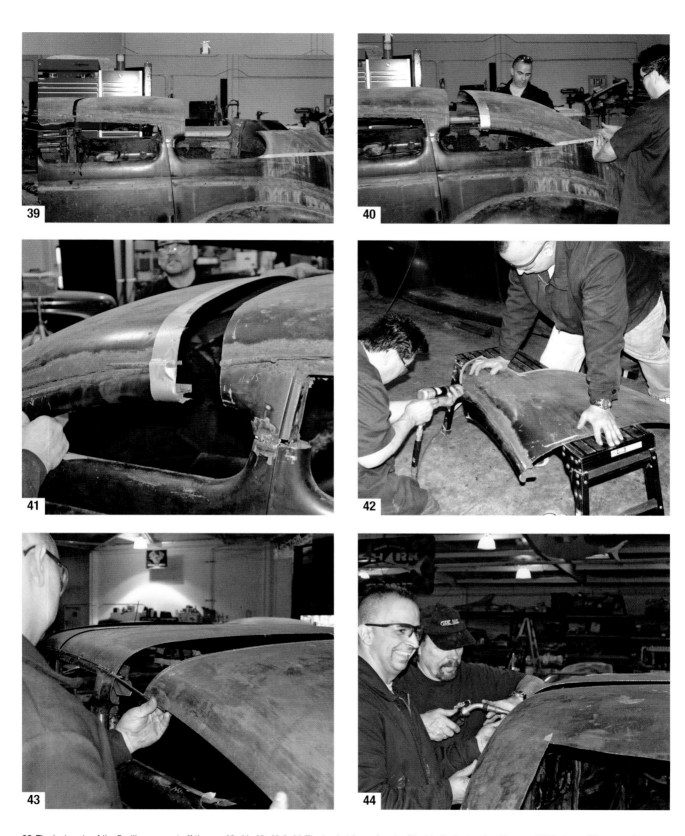

39 The last parts of the D pillars are cut off the car. **40, 41, 42, 43 & 44** The top is trimmed and refitted to the body a few times until it looks as if it cannot fit any better, then it is tack welded into place.

45 *The rear upper window frame is now fitted and welded in place as the top chop nears completion.* **46** *Now the hard part is the making of all the small sections that fill the gaps across the roof. These sections can be made and welded into place one piece at a time.*

47 & 48 *The rear section of the top is missing some metal after the chop, so we measure the area, make a pattern, and fabricate the required piece.* **49** *The lower gutter piece is bent to the required profile and welded in. The rest of the gap across the top is just a single layer of metal and very easy to make.*

50 *I had to cut the rear roof section from front to rear and take out about ¾ inch to get rid of the bulge that was a result of the top narrowing as it was brought down and into contact with the lower body.* **51** *The rear of the roof is also tack welded together along the seam.*

52 & 53 *The roof is also supported by temporary bracing made for just this application.* **54** *The rear section is now welded up using a Mig welder and a hammer for holding the high spots down as we weld. The back window will be filled in with metal after the welding begins.*

CHAPTER 11
TRAINING AIDS

Videos, magazines, and books are the best training aids you can get if you want to learn how to do something. Books are better than magazines, since they have more space, but if you can see something being done in real life it is always much better. We are currently producing an internet show called www.DeadlineTV.net, which is designed to teach all kinds of fabrication skills with all kinds of materials. The show includes interviews with well-known fabricators and, in many cases, we will learn some of their best-kept shop secrets. DeadlineTV.net is free and can be downloaded at your leisure and watched as many times as you want.

We also sell a series of videos that cover a host of how-to topics, such as chopping a top, pinstriping, Frenching headlights, and many other subjects.

Most training aids, such as this book and others like it, are not designed to take you step by step through each process of a subject. They are designed with broader strokes in mind and assume that you have some basic skills already, such as how to adjust an oxy-acetylene torch or how to use a spray gun. Because of this I have found that many basics are never easily learned or taught except by watching a video on the subject.

You can show a picture of a flame coming out of a cutting torch, but unless you see the speed at which it moves when cutting, and unless you hear the sound of a good cut, you are like a one-legged dancer at a waltz—you are missing a lot.

Go to other shops and watch people do whatever it is you want to do, whether that is basic bodywork, metal fabrication, or just welding (but be sure to bring your welding mask). I try to go to other shops just to see the equipment they have and the way they use it. There are shops all around, and some will let you come in and watch. Of course, there are shops that won't allow visitors, but the ones that do are the best anyway because they are proud of what they do and the way they do it.

Watch others who have more skill than you do. Ask questions, but don't just talk. This is a problem I often see.

Someone with little knowledge comes into a shop to ask the owner a question about how to do something and, before anyone can answer, starts telling everyone how he would do what he has just asked how to do.

The point being if you ask an expert and then listen to an expert, you can gain a lot of knowledge. Many people just don't listen, and they wonder why they have problems.

Online forums, such as Metalmeet and OlSkoolrodz, are also good for picking up tricks and techniques and special tools that you can get or even build. One of my favorites is metalmeet.com because as the skill level is way up there. I am amazed at the projects being worked on, such as scratch-building a Bugatti.

Go to the bookstore and pick up books on each subject you want to learn about and then chalk up the cost to education. I love books and have thousands of them on every subject you can imagine, from metal finishing to quantum mechanics. I have found that you can learn the basics of almost any subject within a few days by reading about it. Now you notice I did not say you would be an expert on the subject, I just said that you would learn the basics. To become an expert you will need to spend many years practicing it, doing it until it becomes second nature.

If I want to learn something, I first get magazines and books on the subject and learn the vocabulary of the trade so I can at least "talk the talk." Then I proceed to websites and forums and ask the proper questions. Next, if there are videos on the subject I watch them and take notes. Finally, I compare all the bits of information I've gathered against each other until I have a comprehensive knowledge of whatever it is I want to learn. Last, I apply that new-found information to the trade in which it applies, such as metalwork.

There is a big shortage of good training material on the market, and I run into it all the time when I am doing research on upcoming projects, so the concept of making your own idea is not that abstract. You can also make a good living if you have a few subjects.

CHAPTER 12
MANUFACTURER SOURCE GUIDE

3M Company
(abrasives, paint prep, and metal treatment)
Website: www.3M.com

Ajax Tool Works
(air chisels, hand and power tool accessories)
10801 Franklin Avenue
Franklin Park, IL 60131
Phone: 800-323-9129
Website: www.ajaxtools.com

Auto Body Toolmarket
(auto body and paint supplies, tools)
Phone: 800-382-1200
Website: www.autobodytoolmart.com

Back-A-Line, Inc.
(knee pads, back support belts)
644 11th Avenue
San Francisco, CA 94118
Phone: 800-905-2225
Website: www.backaline.com

Bend-Pak
(automobile lifts)
1645 Lemonwood Drive
Santa Paula, CA USA 93060
Local phone: 805-933-9970, toll-free phone: 800-253-2363
Website: www.bendpak.com

Bessey Tools North America
(clamps, snips, hammers)
1165 Franklin Blvd., Unit G / P.O. Box 490
Cambridge, ON N1R 5V5
Phone: 519-621-7240
Website: www.americanclamping.com

Bosch Tools and Accessories
(cordless and electric power tools)
Phone: 877-BOSCH-99
Website: www.boschtools.com

Channellock, Inc.
(pliers, cutters)
1306 South Main Street
Meadville, PA 16335
Phone: 800-724-3018
Website: www.channellock.com

Chicago Pneumatic
(pneumatic power tools)
Website: www.chicagopneumatic.com

Craftsman
(power and hand tools, tool storage, air compressors)
Website: www.sears.com

Customs by Eddie Paul
(design, custom cars, motorcycles, film/television vehicles, paint and fabrication work, CNC machining, video production)
2305 Utah Avenue
El Segundo, California 90245
Phone: 310-643-8515

Devilbiss Automotive Refinishing
(painting equipment, spray guns, air regulators and filters)
Toll-free phone: 800-445-3988
Website: www.autorefinishdevilbiss.com

Eagle Bending Machines
(hydraulic tube rollers)
Phone: 251-937-0947
Website: www.eaglebendingmachines.com

E.P. Industries, Inc.
(metal fabrication tools, how-to videos/DVDs)
2305 Utah Avenue
El Segundo, CA 90245
Phone: 310-245-8515
Website: www.epindustries.com
www.DeadlineTV.net

ESAB
(welding equipment and consumables)
Website: www.esab.com

Eercoat
(body fillers, primers, metal treatment, fiberglass materials)
6600 Cornell Road
Cincinnati, OH 45242
Phone: 513-489-7600
Website: www.evercoat.com

House of Kolor
(custom automotive paint)
210 Crosby Street
Picayune, Mississippi 39466
Phone: 601-798-4229
Website: www.houseofkolor.com

Hutchins Manufacturing Company
(pneumatic sanders and accessories)
49 North Lotus Avenue
Pasadena, CA 91107
Phone: 626-792-8211
Website: www.hutchinsmfg.com

Ingersoll Rand
(pneumatic power tools)
Website: www.irtools.com

INNOVA—Emissive Energy Corp.
(L.E.D. flashlights)
135 Circuit Drive
N. Kingstown, RI 02852
Phone: 401-294-2030
Website: www.innova.com

Irwin Industrial Tools
(Vise Grips)
Website: www.irwin.com

Lincoln Electric Company
(welding equipment)
22801 St. Clair Avenue
Cleveland, OH 44117
Phone: 216-481-8100
Website: www.lincolnelectric.com

Meguiar's Inc.
(car care products, buffing and polishing compounds)
17991 Mitchell South
Irvine, CA 92614
Phone: 800-347-5700
Website: www.meguiars.com

Milwaukee Electric Tool Corp.
(cordless and electric power tools)
13135 W. Lisbon Road
Brookfield, WI 53005
Toll-free phone: 800-729-3878
Website: www.milwaukeetool.com

Morgan Manufacturing, Inc.
(Morgan "Nokker" slide hammers, accessories)
521 2nd Street
Petaluma, CA 94952
Phone: 800-423-4692
Website: www.morganmfg.com

Motor Guard Corporation
(Magna Spot welders)
580 Carnegie Street
Manteca, CA 95337
Phone: 209-239-9191
Website: www.motorguard.com

National Detroit, Inc.
(pneumatic sanding, grinding, and buffing tools)
P.O. Box 2285
Rockford, IL 61131
Phone: 815-877-4041
Website: www.nationaldetroit.com

P.B.E. Specialities
(paint and body equipment)
13801 Kolter Road
Spencerville, OH 45887
Phone: 888-997-7416

Plasmacam, Inc.
(CNC plasma-cutting machines)
P.O. Box 19818
Colorado City, CO 81019
Phone: 719-676-2700
Website: www.plasmacam.com

POR-15, Inc.
(rust preventative coatings and chemicals)
P.O. 1235
Morristown, NJ 07962
Phone: 800-726-0459
Website: www.por15.com

PPG Industries (factory and custom automotive paint)
Website: www.ppgrefinish.com

Ranger Products (floor jacks, stands, and tool storage)
1645 Lemonwood Drive
Santa Paula, CA USA 93060
Local phone: 805-933-9970, toll-free phone: 800-253-2363
Website: www.rangerproducts.com

Ringers Gloves (work gloves, shoes)
335 Science Drive
Moorpark, CA 93021
Phone: 800-421-8454
Website: www.ringersgloves.com

Sata (automotive painting equipment)
Website: www.sata.com/usa/

Scotchman Industries (metal fabrication equipment)
180 E. Hwy 14
P.O. Box 850
Philip, SD 57567
Phone: 800-843-8844
Website: www.scotchman.com

SEM Products, Inc. (custom paint, prep and repair products)
651 Michael Wylie Drive
Charlotte, NC 28217
Phone: 1-800-831-1122
Website: www.sem.ws

Sharpe Manufacturing Company (spray guns and accessories)
P.O. Box 1441
Minneapolis, MN 55440
Phone: 800-742-7731
Website: www.sharpe1.com

Slidge Sledge (sliding hammers)
2500 W. Higgins Road
Hoffman Estates, IL 60195
Phone: 800-276-0311
Website: www.slidesledge.com

Snap-On Tools (automotive tools and tool storage)
Website: www.snapon.com

Tools USA—Standard Tools and Equipment
(automotive spray booths, automotive tools)
4810 Clover Road
Greensboro, NC 27405
Toll-free phone: 800-451-2425
Website: www.toolsusa.com

Tru-Line Laser Alignment
8231 Blaine Road
Blaine, WA 98230
Phone: 800-496-3777
Website: www.tru-line.net

INDEX

All Metal, 89

aluminum alloys, types, 8, 9

annealing
  metal, with planishing
    hammer, 49, 50
  sheet aluminum, 14–16
  sheet steel, 13, 14

AutoCad, 117–123

Automotive Classics, 89

Bend Pak MD-6XP Scissor
  Lift, 73

Bend Pak/Ranger floor jack, 70

bending tubes, 79–81

Bessey
  aviation snips, 58
  Clamps, 70
  magnetic tools, 59

Beverly shear, 43

Bosch Tools
  Blue Core battery technology, 57
  cordless saw, 57
  Model 1521 shear, 56
  Model 1662B circular saw, 57
  Model 1873-8 grinder, 57
  multiple bay charger, 58
  nibblers, 57

Boss Hoss, 15

Cartesian coordinates, 75, 76

Channellock Code Blue pliers, 60

Chicago Pneumatic
  air ratchet, 62
  CPA7300 drill, 62
  CPA835 nibbler, 57
  Pneumatic grinder, 59, 61
  polisher, 64
  reciprocating saw, 61
  sander, 61, 64
  saw, 55

CNC (Computer Numeric
  Control), 74, 75

Cobra movie car, 10, 82

CP palm sander, 63

CP sander, 64

Craftsman air compressor, 68

Craftsman drill press, 68

Customs by Eddie Paul, 8, 59
  sheet metal roller, 71
  English Wheel, 46

D-All vertical band saw, 72

DeadlineTV,net, 156

Descartes, Rene, 76

DeWalt drill/driver, 55

Discovery Channel, 10, 37

E.P. Industries, 49, 60

English wheel, 71
  Monster Wheel, 66
  planishing hammer, 71
  tube roller, 80

Eastwood Company
  heat lamp, 54
  Heavy-Duty Anti Rust, 98, 99
  manual shrinker and s
    tretcher, 71
  plastic prying tools, 53
  PRE, 90
  tubing roller, 60
  read roller, 70

Eco-Jet, 36

English wheel, 46–51
  basic tracking, 49
  features and descriptions,
    46–48
  overview, 46
  practicing, 48, 49
  tips, general, 47, 48
  tools to use along with, 50, 51

error, accumulation, pattern
  making, 17, 18

Fadal CNC vertical machining
  center, 76

firewall, building, process, 82–88

floorboard, making,
  process, 18–34

Grant steering wheel, 70

hammer forming, relevant
definitions, 35, 36

hardening sheet metal, 13

headlights, Frenched, fabricating,
  124–133

Herkules paint-gun washer, 89

hood, extending, 136–139

Hutchins Mfg. sander, 54

Ingersoll Rand, 43
  air ratchet, 62
  belt sander, 63
  polisher, 64
  sander, 64

Irwin
  aviation snips, 58
  clamps, 63
  pliers, 64
  tool caddy, 65
  Vise Grip locking pliers,
    62–64, 73

Jelutong, 35, 36, 38

Jet band saw, 72

K2 CNC machine, 54

Leno, Jay, 36

M&K Metal, 11

manufacturers, list of, 157, 158

MBX Power Tool, 66

metal
  finishing, overview, 89
  repair, basic, 114–116
  skirt, making, 134, 135
  repairing in, 90

metallurgy fundamentals,
  overview, 7, 8

Milwaukee Electric Tools shear, 67

Mittler Brothers
  Bend Calculator, 101
  hydraulic bending machine, 69
  protractor, 103
  shoe and follow bar, 104
  Smart Tool Angle Sensor, 103
  tube bender, 102, 103

mockups, pattern making, 18

Morgan Manufacturing
  Nokker, 53

National Detroit sander, 54

NBC Universal, 36

planishing hammer, 49–51

plug, making, for hammer
  forming, 36–38

POR-15 Metal-Ready, 93, 96

Powermaxx 1000 Hypertherm, 53

project considerations, 108, 109

Proto Trak VM, 76

rear end, four-link, fabricating, 117

Ringer Gloves, 66

roll bar fabrication, selecting parts
  for, 100, 101

rust, 91–99
  overview, 91, 92
  prevention, 92
  treating, 92, 93
  treating, process, 94–99

Scotchman, 104
  belt notcher, 102
  tube notcher, 56, 81

Secret Weapon, 78

sheet metal qualities, 8

shrinking sheet metal
  considerations, 12
  process, 51, 52

softening sheet metal, 13

square tube, 104

steel, types, 9–11

stress, sheet metal, defined, 8

stretching sheet metal
  considerations, 12, 13
  process, 51, 52

taillights, sunken, fabricating, 113

tensile strength, sheet metal
  defined, 8

thickness, sheet metal, 11, 12

Tightbond glue, 37

tools
  bending, 79–82
  cutting, 77, 78
  metalworking, overview, 53–73

top, chopping, process, 144–155

training aids, 156

trunk panel, interior, aluminum,
  making, 110–112

tube bend calculator, 101

tube bending
  machines, 79, 102
  measuring for, 101
  process overview, 104–107
  techniques, 102, 103

tube notchers, 81

tube-rolling machines, 79

Vivex, 32, 41

Weldwood glue, 37

wheel flares, making, 38–45

windshield divider, split-front,
  process, 140–143

xXx, 100

yield point, sheet metal, defined, 8

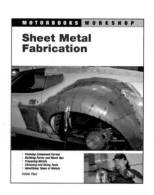 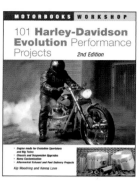  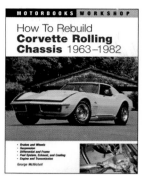